The Role of Oxygen in Improving Chemical Processes

The Role of Oxygen in Improving Chemical Processes

Edited by

M. Fetizon
École Polytechnique, Palaiseau, France

W. J. Thomas
University of Bath, Bath, UK

ROYAL
SOCIETY OF
CHEMISTRY

The Proceedings of the 6th BOC Priestley Conference, sponsored by
BOC Ltd and organized by The Royal Society of Chemistry,
in association with the Société de Chimie Industrielle and the
Société Française de Chimie in Paris, France, 7–9 September 1992.

Special Publication No. 132

ISBN 0-85186-725-1

A catalogue record for this book is available from the British Library

Published by The Royal Society of Chemistry,
Thomas Graham House, Science Park, Cambridge
CB4 4WF

Printed by Hartnolls Ltd, Bodmin

Preface

BOC Priestley Conferences, organised by the Royal Society of Chemistry and sponsored by BOC Ltd, have been a feature of the strong links between academe and industry since the first of these triennial conferences held at Leeds in 1977. The theme of each of the five conferences until 1989 has been to commemorate the life and work of Joseph Priestley, the discoverer of oxygen and some of the rare gases, dissenter from many of the established practices and ideals in eighteenth century Britain, and enthusiastic exponent of the burgeoning Unitarian theology.

Priestley was a philosopher and *éminence gris extraordinaire*. Some of his scientific work, especially that concerned with oxygen, was associated with the work of Lavoisier. On the one hand Priestley enunciated the phlogiston theory of combustion, whereas Lavoisier preferred his acidifying principle (*principe oxygine*). Much debate has ensued concerning the credit which should be attributed to one or the other. As with so many scientific discoveries, the truth probably lies in the not uncommon phenomenon that ideas and theories are developed through discussion and correspondence. The first to make a discovery often becomes obscured by the gradual development of the theory as new facts are added and the theory is consequentially modified. What is not controversial is the undoubted scientific contribution which both Priestley and Lavoisier provided to the advancement of fundamental chemical knowledge.

In this year of 1992 immediately preceding the formation of the European Common Market when Britain and France, two members of the European Community, are consolidating their common purpose, it is singularly appropriate that this sixth BOC Priestley conference was held at the Maison de la Chemie in Paris and jointly organised by the Royal Society of Chemistry, the Societé de Chemie Industrielle and the Societé Français de Chemie. Some departure from the format of the previous conferences therefore seemed appropriate. The organising committee decided to include a round-table discussion following the conclusion of the two sessions concerned with combustion processes and to have two, rather than the usual one, schools' competitions encouraging young aspiring scientists to produce a painting or collage representing the life of Priestley (British schoolchildren) and Lavoisier (French schoolchildren).

Papers in this volume were those invited by the scientific programme committee chaired by M Daniel Deloche of L'Air Liquide. Themes included sessions on combustion processes, chemical synthesis, wet oxidation and biological processes. Each of these themes either directly or indirectly involved oxygen and oxygen transfer within the topic presented. Fourteen such papers are included in this volume. A round-table discussion followed one of the sessions on combustion processes and a transcript of this discussion is duly recorded. Although not contained in this publication, ample discussion followed each session and the scientific programme was strongly supported by the independent display of eighteen posters.

Two recurring features of the Priestley conference are the Priestley Lecture and the BOC Centenary Lecture. The former was delivered by Hubert Reeves (of CEN Saclay) who traced how the stars synthesised oxygen through nuclear reactions. No written paper was available to describe this lecture but a transcript of the lecture by Mr Ray Corns is included in this volume. Chris Wright of AEA Technology delivered the BOC Centenary Lecture which described how the use of oxygen can reduce pollution and improve the environment.

Also included in the conference programme was a session devoted to the history of pollution and the gradual development of skills to protect the environment from a variety of potential sources of contamination. Thus the atmosphere, industrial waste, the water industry and food were topics presented in papers which sought to trace the development and application of science to such problems requiring solutions, many of which involved oxygen as the ameliorating antidote. Seven papers were presented in this session which has been a regular feature of previous BOC Priestley conferences.

It is hoped that the present collection of papers will be of interest to many in academe and industry, to chemists and chemical engineers, and to technologists working in a broad spectrum of processing where oxygen is a prime constituent of the process. It is evident at least that the themes have proved unifying across a number of disciplines.

W J Thomas
Editor

Contents

BOC Centenary Lecture 1992

Improving on Nature with Oxygen

C. J. Wright

AEA TECHNOLOGY, HARWELL LABORATORY, OXFORDSHIRE
OXI I ORA, UK

The title that I've chosen for this lecture needs some explanation.   What I have in mind and what I wish to talk about is the use of oxygen to improve the environment or, expressed more precisely, the use of oxygen to minimise the impact of man's activities on other species.   Oxygen makes a major contribution to reducing environmental pollution and the purpose of my talk is to survey those areas where its application has most impact.   I will not be talking about the use of oxygen to accelerate biological processes as in industrial fermentations.   There was no intentional hubris in my title and I hope the gods will pardon any suspicion of arrogance on my part in suggesting that it would be possible to improve on the natural world.

I will be covering in my talk a number of sub-themes, many of which will be dealt with in more detail and by people who are more knowledgeable and more expert than myself later in this Conference.   I will be describing how oxygen is used for the treatment of different forms of waste: for gaseous emissions, for liquid effluents and for solids.

Figure 1 is a table of those areas where oxygen has the greatest impact in waste treatment technologies classified into current and future major waste treatment technologies.

Throughout this talk I want to reinforce the message that oxygen technologies are not only economically attractive but they can also be environmentally beneficial.   Furthermore that they can have an enabling function for other technologies by rendering other processes more environmentally benign than they would be otherwise.   Before I proceed too far in this direction, however, we should recognise that there is an environmental cost of producing and transporting oxygen and we should satisfy ourselves that these costs are small in relation to any benefits that might arise from oxygen use.

The major environmental costs of oxygen production, ignoring the questions of the aesthetics of power stations and accidental releases of CFCs from refrigeration cycles, are related to the masses of acid gases that are a by-product of electricity generation from fossil fuels.   Taking values for the UK, which is not an atypical economy in terms of fuel mix, and taking values of the masses of acid gases produced per unit of electricity generated and combining these numbers with typical data on the efficiency of generating liquid oxygen, we find the indicative numbers shown in Figure 2.

FIGURE 1

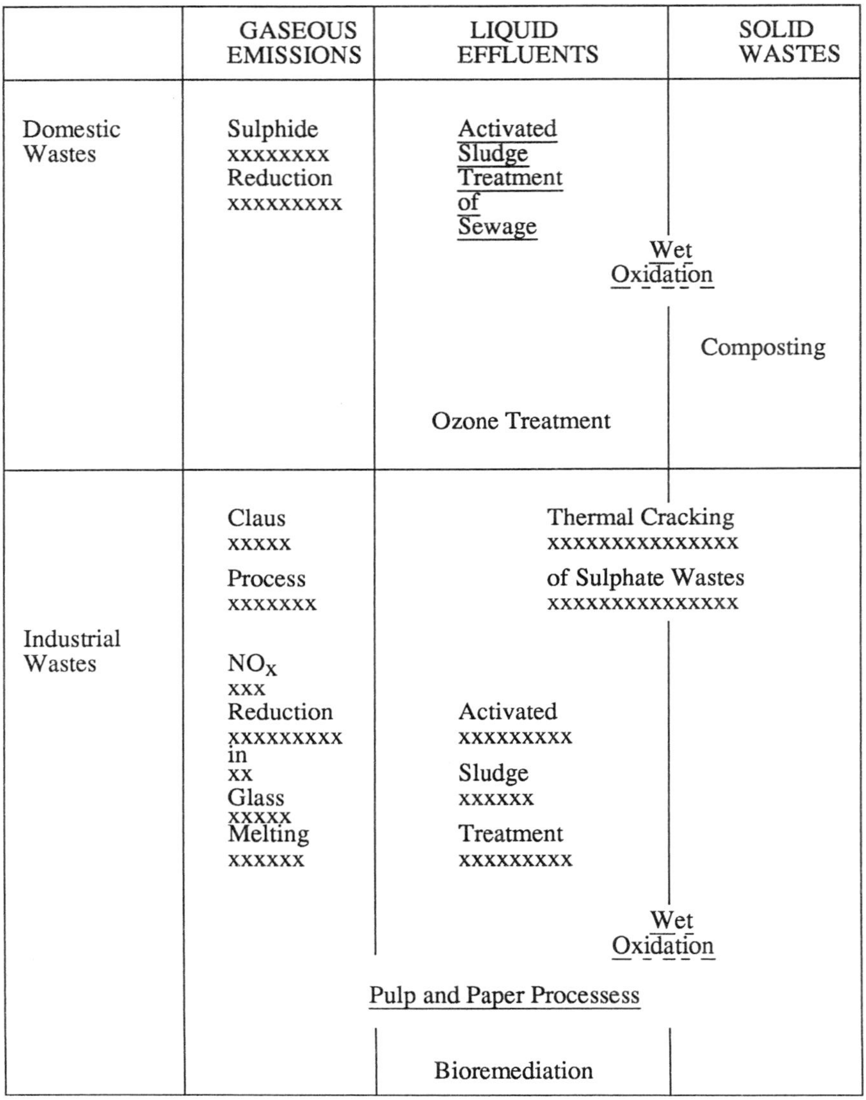

	GASEOUS EMISSIONS	LIQUID EFFLUENTS	SOLID WASTES
Domestic Wastes	Sulphide xxxxxxxx Reduction xxxxxxxxx	Activated Sludge Treatment of Sewage Ozone Treatment	Wet Oxidation Composting
Industrial Wastes	Claus xxxxx Process xxxxxxx NO$_X$ xxx Reduction xxxxxxxxx in xx Glass xxxxx Melting xxxxxx	Thermal Cracking xxxxxxxxxxxxxx of Sulphate Wastes xxxxxxxxxxxxxx Activated xxxxxxxxx Sludge xxxxxx Treatment xxxxxxxxx Wet Oxidation Pulp and Paper Processess Bioremediation	

Current Major Users ____   Current Important Users xxxxxx

Future Major Users? - - - -

FIGURE 2

ACID GAS PRODUCTION ASSOCIATED
WITH LIQUID OXYGEN PRODUCTION

ACID GAS	MASS OF ACID GAS/ MASS OF OXYGEN PRODUCED
Sulphur Dioxide	6.7×10^{-3}
Oxides of Nitrogen	1.9×10^{-3}
Carbon Dioxide	0.49

These numbers ignore the smaller quantities of acid gases produced in distributing liquid oxygen.

There is not yet, to my knowledge, any measure of environmental impact cost for different chemicals which would enable one to compare for instance the cost of an equivalent of CO_2 generated with an equivalent of NO_x avoided but nonetheless, where direct intercomparisons on the same gas can be made the benefits of using oxygen can outweigh the costs by many orders of magnitude. A good example is sulphuric acid waste recycling which I'll be discussing later in this talk.

Another example is glass melting where the use of oxy-fuel leads to a significant reduction in the total mass of NO_x generated per mass of glass melted. However we must note that both German and U.K. legislation defines emission limits, rightly or wrongly, in terms of the concentration of NO_x emitted in the flue gas and in these terms the benefits of oxy-fuel are obviously less striking because of the smaller volumes of waste gas.

I also want to make the additional point that the NO_x emissions in the production of oxygen are of a similar order as the NO_x emissions that the glass industry is seeking to control by the increasing utilisation of oxygen. The figure in my table is equivalent to 2.1 gms NO_x/M^3 O_2 where as the TA LUFT figure for NO_x limits is $1.2 \rightarrow 3.5$ gms NO_x /M^3 of reference off-gas.

Gaseous Emission Technologies

In this area oxygen finds important applications in the control of discharges both of NO_x, which I've just alluded to, and of volatile sulphur compounds, three examples of which are the suppression of anaerobic fermentation in sewage pipelines by direct injection of oxygen, the enhancement of the Claus Process, and black liquor oxidation in wood pulp making. I'll say more about black liquor oxidations later in my presentation and concentrate for the moment on the first two examples.

In sewage pipelines of a certain design the pipeline conditions can become anaerobic and anaerobic fermentation takes place. This leads subsequently to the generation and release of hydrogen sulphide and thiols which creates offensive odours and sour gas corrosion of the pipelines. A simple solution to the problem is the injection of oxygen to maintain aerobic conditions which then keeps everything smelling sweetly.

The second example where sulphur emissions occur and where oxygen is used is in refineries where the Claus Process is the name given to the process used for the conversion of hydrogen sulphide to sulphur. It consists of a combustion step in which part of the hydrogen sulphide is converted to SO_2 and a precipitation step in which the other part of the hydrogen sulphide stream is reacted catalytically with the sulphur dioxide from reaction (1) to produce sulphur.

FIGURE 3

$$2H_2S + 3O_2 \rightarrow 2SO_2 + 2H_2O \quad (1)$$

$$2H_2S + SO_2 \rightarrow 3S + 2H_2O$$

As the sulphur content of refinery feedstocks rises (which is what is predicted) and as the sulphur contents of refinery products reduce in response to regulations, then the requirement for increased sulphur removal i.e. Claus capacity, increases. Oxygen can be used to augment the capacity of existing Claus plant by overcoming throughput limitations in the combustion stage of the process and oxygen enrichment or pure oxygen combustion processes can both be retrofitted to existing plant. In the latter case, the extra heat release is likely to lead to temperatures in the furnace which could exceed the safe limit for the refractories and a number of proprietary staged combustion, combustion with moderators or recycle technologies have been described in order to overcome this problem[1] (Fig. 4). In the last five years there have been a number of conversions to oxygen technology in the US, Europe and Japan and, as regulations bite and feedstock prices change, the application is likely to become increasingly widespread.

FIGURE 4

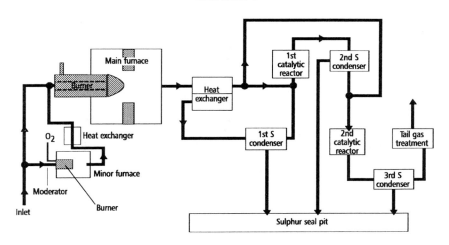

One emission control technology that I'd like to mention briefly, primarily because of the worldwide interest that is being shown in it at the moment, is the use of microwave or r.f. air plasmas for the elimination of low concentrations of volatile organic hydrocarbons. The elimination of low level V.O.C. emissions is not an easy problem and plasma technology is a novel and conceptually straightforward solution which works by generating high concentrations of oxygen anions and radicals at ambient pressures and temperatures. The high reactivity of these species compensates for the low concentrations of the V.O.Cs.

Liquid Effluent Technologies

Probably the most widespread source of liquid effluent is domestic sewage and probably the most widespread environmental application of oxygen is the sewage treatment process, oxygen-enhanced, activated sludge treatment. In this section of my talk I want to describe the activated sludge process and some of its applications including those outside domestic sewage treatment, together with two other less common but related technologies, wet oxidation and the still-to-be commercialised oxygen enhanced composting. Secondly, I also want to spend a few minutes touching on the industry-specific, liquid effluent treatment problems found in the pulp and paper industry.

Oxygen-enhanced, activated sludge treatment is the process of injecting oxygen into waste water treatment tanks. In these tanks naturally occurring microorganisms, known as activated sludge, digest the organic materials and grow in mass as a consequence. The sludge is then separated from the treated water which can be released back into the environment. The digestion process is oxygen limited under normal conditions, a limitation which is overcome by the use of aerators in conventional technology. Higher concentrations of oxygen and therefore higher digestion rates can be obtained with pure oxygen injection which needs to be introduced as small bubbles and to be widely dispersed for high oxygen utilisation efficiencies. Figure 5 shows a commonly used method for oxygen treatment with sidestream, venturi and oxygen injector.

FIGURE 5

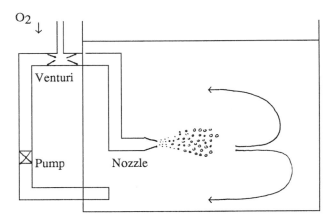

Biological wastes such as effluent from vegetable processing plants, together with selected types of non-toxic chemical wastes such as coke-oven effluents, and wastes from textile and speciality chemical plants, are commonly treated in this way.   The benefits of this intensification technology are primarily those of capital cost reduction and a lower requirement for land.   The process,   is well known.   It is one of the largest applications for liquid oxygen in Western Europe and will be the subject of a paper later in this Conference by Professor Smith.

The by-product of this process, which is in effect a method of concentrating biological wastes, is the growth of the activated sludge which also needs to be disposed of after separation from the treated waste.   Historically these wastes have been disposed of by being spread on the land, by being dumped into landfill sites and, historically, in the case of the UK by being disposed of at sea. Sea dumping is no longer acceptable and in many countries there is an increasing shortage of suitable sites for the disposal of sludge on land.   Because of the high concentration of pathogens in sewage sludge, the land on which the sludge has been spread cannot be used for alternative purposes for many years.   As a consequence, attention has been turned to incineration and the competitive technology of wet oxidation.

Wet oxidation is the high-pressure, autothermal, oxidation of organic wastes in water and because of its many intrinsic attractions, considerable sums have been invested in its development.   Very simply, the technology involves the reaction of oxygen at high pressure (many tens of atmosphere) and high temperature (in excess of 200°C) with hydrocarbon wastes dispersed in water.

FIGURE 6

Hydrocarbon sludge + O_2 $\quad\quad$ $\dfrac{>200°C}{>10\text{Ats}}$ $\quad$ $CO_2 +$ $\quad$ H_2O

Under these conditions, the oxidation of the water proceeds almost to completion and sewage sludge, for instance, is converted into CO_2, H_2O and inorganic salts.

When compared with other processes for waste oxidation, wet oxidation has the following advantages:

FIGURE 7

ADVANTAGES OF WET OXIDATION

Efficient in the use of land because of the high rate of reaction.

Readily controllable because of the thermal mass of the water and the low concentration of reactants.

Capable of proceeding almost to completion so avoiding the need for further processing.

Avoids potentially harmful by-products

Capable of treating a wide variety of different wastes under different conditions.

A number of proprietary wet oxidation processes exist of which those listed in Figure 8 make use of "pure" oxygen rather than air as the oxidant. One of these processes, the Modar Process, operates above the critical point of water (374°C and 226 Bar). At these high temperatures and pressures, water behaves as a non-polar solvent and high oxidation efficiencies of mixtures such as PCB contaminated transformer oil and chlorphenol/nitrobenzene mixtures have been reported.[2] Furthermore, inorganic contaminants, including toxic metals, are insoluble under these conditions and are readily separated from the reactants.

The other processes operate at subcritical pressures but still under conditions where the solubility in water of most hydrocarbons is considerably enhanced.

FIGURE 8

PROPRIETARY WET OXIDATION PROCESSES

Process	Temperature °C	Pressure Bar	Retention Time Mins	Comments
Vertech	175 → 325	70 → 160		Sub-surface Exploits Drilling Technology
Zimpro	175 → 325	80 → 210	20 → 240	Surface Mounted
Modar	700	250	~1	Supercritical Surface Mounted
Kenox	200 → 250	40 → 50	60	Air Based

The Vertech process, Fig. 9, achieves its chosen pressure hydrostatically by utilizing drilling technology to conduct the reactants to the bottom of a very deep well.

FIGURE 9

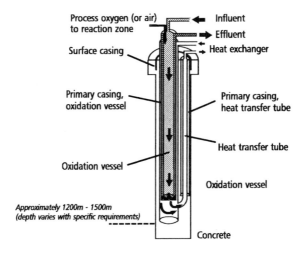

Process oxygen (or air) to reaction zone
Surface casing
Primary casing, oxidation vessel
Oxidation vessel
Approximately 1200m - 1500m (depth varies with specific requirements)
Influent
Effluent
Heat exchanger
Primary casing, heat transfer tube
Heat transfer tube
Oxidation vessel
Concrete

The limited scale, to date, of demonstration projects for some of these technologies has meant that there are appreciable uncertainties associated with their "in-service" costs.   This is particularly so with the Vertech and Modar processes, where because of their design there is interest in the long term costs associated with abrasion and corrosion of the reactor vessels.   The commissioning of the Vertech reactor at Appeldoorn in the Netherlands will go a long way to answering many of these questions.

Nonetheless, despite these un-certainties and questions about how maintenance could be conducted in a down hole reactor, independent assessments of lifetime costs suggest that wet oxidation is financially attractive when compared with incineration and other disposal processes.   It seems likely therefore that they will be used increasingly as experience leads to greater confidence.

An alternative process for sludge disposal, even less developed than wet oxidation, is the process of oxygen-enhanced composting.   In conventional composting technology sewage sludge together with leavening agents such as wood chips and straw are allowed to autodigest.   The final product is a fertile material, for which there is a proposed EC specification, which can be be spread on to land and used as a soil improver.

The autodigestion proceeds in two stages via aerobic and then anaerobic fermentation.   During the first stage the compost is heaped into piles and subject to forced ventilation to maintain temperatures between 40 and 65°C. Whilst this process proceeds with air ventilation at a favourable rate in equatorial climates, in Northern Europe oxygen injection is needed if the process is to proceed at an acceptable rate.   Without oxygen the heat of combustion will not compensate for the evaporative heat losses associated with the forced ventilation.   Conversely too much oxygen at this stage will raise the temperature above 65°C, so destroying the micro-organisms which can break down long-chain polymeric material such as cellulose and lignin.   The aerobic fermentation stage proceeds for about fourteen to twenty-eight days after which time the oxygen supply is discontinued and secondary, anaerobic fermentation is allowed to proceed for up to ~ three months.

In contrast with wet oxidation the composting process is lengthy and requires considerable areas of land.   Furthermore the end product cannot be cheaply returned to the land if the sludge has been contaminated with heavy metals. On the other hand the process is of low capital cost and maintenance free and could find application in areas of low land costs.

Another process of interest is the high-pressure ozone treatment of sewage sludge, a process which is in use at one site in the U.S. (West New York) and which has been demonstrated in the UK.   Its prime advantage is that it provides a speedy route to reducing, by many orders of magnitude, the concentrations of pathogens in sewage sludge.   Thus sludge which has been treated in this way can be spread on land and the land can be returned to use, as pasture say, after a much shorter timescale than would alternatively be the case.   The process [3] is quite simple involving the passage of oxygen through an ozone generator before a 3% ozone in oxygen mixture is introduced with the sludge into a high pressure (about 4lbs) stirred tank reactor.

For the sake of completion it is also worth making a brief mention of ground decontamination technology specifically the removal of hydrocarbons from contaminated soil.   There is the brute force method of total incineration of the contaminated mass and the more subtle, but complex, approach of in-situ remediation.   Remediation technologies using oxygenated water have been demonstrated but they face two problems.   The first is the problem of oxygen solubility for where oxygen concentration is the rate limiting parameter, then 25kg of water are required to remove 1gm of hydrocarbon from soil, assuming that all the water is saturated.   In contrast, 4kg of water saturated with H_2O_2 would suffice or only 13 gms of air, if gas phase oxidation rather than liquid phase was the preferred choice.   A second problem concerns the experimental observation that biodegradation with dissolved oxidants has only a partial impact on the concentrations of insoluble hydrocarbons in soil. For the moment it appears that simpler technologies, Fig. 10, such as soil venting, which are being used in large scale tasks by the US Air Force,   are likely to have the edge over dissolved oxygen technologies.

FIGURE 10

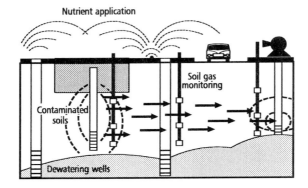

Pulp and Paper Effluents

I have dedicated a section of this presentation to pulp and paper effluents because of the importance of oxygen in reducing the environmental impact of processes which would otherwise be significant polluters and because of the uniqueness of some of the process technology involved.   Furthermore, some of the oxygen applications are attractive because they avoid the formation of effluents rather than help dispose of the wastes of conventional technology. Again, this topic will be dealt with in more depth by Dr. Croon later in the Conference and I will therefore content myself with a few general remarks.

Wood pulp making has the following significant effluent disposal problems:

FIGURE 11

EFFLUENT DISPOSAL PROBLEMS OF WOOD PULP MAKING AND THEIR SOLUTIONS

Problem	Solution
- Sulphite waste liquors from the delignification process	Recycle via combustion process
- Sulphur containing gaseous emissions	Oxidise
- High B.O.D., C.O.D. and colour of bleaching plant effluents	Activated sludge treatment
- Toxic chemical concentrations, chlorine compounds in bleach plant effluent.	Oxygen bleaching

and oxygen technology is able to bring significant benefits in each case.

In a conventional chemical pulping procedure, wood chips are broken down in a digester using what is known as the Kraft or sulphate process. I won't go into the chemistry of this process here apart from saying that the products of the digestor, namely "black liquor" and pulp are then separated so that the pulp can be pumped away for bleaching, whilst the "black liquor" which contains residual NaOH and NaHS, left over from the digestion liquids, is recycled.

The recycling consists of the recovery of the sulphur values and the combustion of the organic material in the black liquor to raise steam in a cycle comprising evaporation; combustion in a recovery furnace, and a dissolver and make-up tanks for the salts extracted from the furnace. In the evaporation process, oxygen is introduced into a packed column in what is known as black liquor oxidation in order to convert residual hydrosulphides and mercaptans present in the black liquor to less volatile sulphur compounds, and so prevent the release of H_2S and mercaptan from the flue gases.

FIGURE 12

$2NaHS + 2O_2$	$\rightarrow$	$Na_2S_2O_3$	$+ H_2O$
$2CH_3SNa + \frac{1}{2}O_2 + H_2O$	$\rightarrow$	CH_3SSCH_3	$+ NaOH$

High process efficiencies are required if emission standards are to be met.

In an integrated oxygen-using process, oxygen is also used in the recovery furnace for enhancing the combustion of the concentrated black liquor [4] (Fig. 13). The liquid is sprayed into a heat recovery boiler where the droplets dry and then combust in the furnace atmosphere.   Reducing conditions in the bottom of the furnace convert the sulphur compounds plus added sulphate to sodium sulphide which runs out of exit ports.   Oxygen introduction in the upper part of the furnace enables either a capacity increase to be obtained from a pre-existing boiler or alternatively the construction of a smaller unit than would be possible in the absence of oxygen.   Even so, oxygen concentrations need to be carefully controlled since strongly oxidising conditions in the upper reaches of the furnace will convert SO_2 to SO_3 which in turn will react with sodium sulpate to form the pyrosulphate which melts at 400°C

$$Na_2SO_4 + SO_3 \rightarrow Na_2S_2O_7$$

leading to a glassy corrosive deposit within the boiler.

The pulp which has been separated from the black liquor is then bleached and in conventional chemical bleaching, low temperature, mild conditions are used in order to avoid deleterious oxidation of the wood carbohydrates.   Conventional oxidising chemicals are chlorine, chlorine dioxide, and hypochlorite which may be used in serial combination, i.e. one after the other in a variety of sequential oxidations separated by alternative washing steps.   Because of the concern associated with the build-up of chlorinated hydrocarbons in the environment, the attractions of oxygen bleaching are increasing as a means of reducing the requirement for chlorine.

FIGURE 13

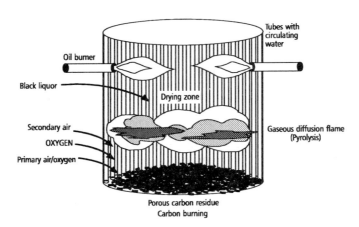

In one common sequence of bleaching processes, oxygen is used as an initial stage prior to chlorination, which is in turn followed twice by a ClO_2 bleaching stage.   Reactions are carried out at ~ 85 $\rightarrow$ 100°C ~ 2 - 3 wt%

NaOH. Another common sequence is to use oxyg:n in a caustic extraction process after an initial chlorine based bleaching step (Figure 14). Empirically it has been shown that oxygenation of some of the liquor that is injected into a conventional extraction tower can lead to significant (15 → 50%) reductions in the consumption of bleaching chemicals.[5] More recent developments include the Prenox process which is claimed to be the most promising method for supressing and even eliminates the use of Cl_2 in Kraft Pulp bleaching. It involves pretreatment of the pulp with NO_2, followed by O_2 bleaching. [6]

FIGURE 14

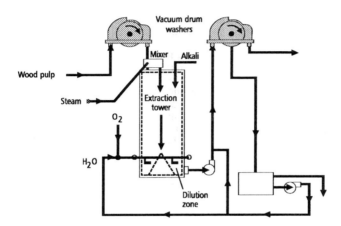

In addition to these processes in which oxygen is substituted either for air or chlorine in the classical chemical process technology, oxygen based improvements to the basic Kraft process have also been developed. One such process involves the white liquor which is the final liquid product arising from the Kraft liquor recycling, i.e. the redissolved sodium hydroxide sulphide and sulphate, prior to being used once again for wood chip digestion. The presence of polysulphides obtained by a sulphur dissolution process in this white liquor increases the yields from the Kraft process by up to ~ 10%. More recently it has been recognised that oxygen oxidation provides a low emission alternative to the sulphur dissolution process for increasing the concentrations of Na_2S_x although the increase in yields are more limited to 2 - 2.5%.

Finally, soda-oxygen pulping can be used as an alternative to the Kraft process, since oxygen itself is a good delignifying agent in alkaline solutions. The difficulties associated with the process are the competing reactions of carbohydrate oxidation and the relatively slow rate of reaction exacerbated by the low available oxygen partial pressures in this diffusion controlled reaction. In this process oxygen is generally introduced into an alkaline wood pulp through a high shear mixer which simultaneously leads to intimate mixing and high oxygen bubble surface areas.

Solid Waste Disposal

Although my classification of wastes into solid, liquid and gas is somewhat arbitrary since many wastes are colloidal dispersions, I have classified my next example as an example of solid waste disposal since it deals with the disposal of ammonium sulphate.

The application is of particular relevance at this BOC Priestley conference in Paris, since the technology is the oxygen combustion of sulphate waste which is being introduced by ICI and Air Liquide at Teesside in the UK and which will use 600 tonnes/day of BOC supplied oxygen.[7]

In the acetone cyanhydrin process for methyl methacrylate production, ammonium sulphate is a byproduct.   This can be either converted into fertilizer grade ammonium sulphate or burnt in air in an acid recovery plant for reconversion to sulphuric acid for recycling.   At ICI's Teesside plant 180,000 tonnes per year of waste are being dispersed into the sea rather than being recycled and when ICI recently decided to expand production of methyl methacrylate by 100%, it also decided to invest in oxygen-based waste combusion at the same time (Fig.15).

FIGURE 15

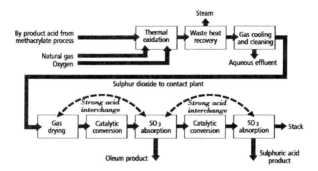

Pure oxygen-based combustion rather than air or enriched air was chosen because of the significantly lower capital costs that result from the reduced volumetric throughputs and because of the advances that have been made in furnace design as a result of computational fluid dynamic modelling.

The risks of developing a burner that fails to meet its specification, or of designing a furnace in which the temperature distribution exceeds the specification of the refactories, have been considerably reduced as a result of the wide availability of CFD.   CFD models of complex dynamical processes involving chemical reactions have become increasingly sophisticated and the leading CFD codes have been widely validated by comparison with experimental observations.   Thus it is possible to design furnaces with lower lower margins of error which has a further and direct impact on capital costs.

Combustion of spent acid wastes from other sources such as petroleum refineries or pickling liquors has been both practised and described in the patent literature.[8]   Apart from these examples, however, instances of oxygen being used exclusively for the disposal of solid wastes are few and far between. The advantages of oxygen use in waste incineration, such as the high rate of

FIGURE 16

Advantages of Oxygen Use in Combustion Processes

> Higher rates of heat transfer
>
> Higher fuel utilisation efficiencies
>
> Higher combustion temperatures
>
> Lower mass flows

heat transfer, have not been found to be generally beneficial except where fluctuations in the calorific value of the material needing incineration need compensating changes in oxygen concentrations in order to maintain the combustion temperature and prevent emissions of particulates and unburnt hydrocarbons. Similarly the high temperatures achievable in oxygen combustion have not as yet found widespread application, despite the consequential benefit of being able to achieve higher levels of destruction of particularly pernicious chemicals at the higher temperatures. In contrast, much experimental effort from the gas industry has been spent in developing burner technology which minimises the potential environmental disadvantages of higher combustion temperatures; namely higher NO_x concentrations.

Nonetheless the above advantages are real advantages and it may be just a matter of time before regulatory pressures overcome conservatism and oxygen use becomes more widespread.

Conclusion

In conclusion I trust that I have shown that oxygen technology has a valuable rôle in reducing the impact of man's activities on his environment. Oxygen technologies have come a long way from the metal cutting and smoke stack industries with which they were once associated. Furthermore, I'm sure that Priestley and Lavoisier would be highly gratified to see the ways in which their gas was being used to improve upon nature.

References

1. R. W. Watson. Eur-Pat-Appl 0237216. 1987.
2. C. N. Staszak, K. C. Malinowski and W. R. Killilea. Env.Prog. 1987. 6, 39.
3. R. N. Edwards.   U.S. Patent 3,772,188.
4. W. T. Mullen.   U.S. Patent 4,857,282. 1989.
5. R. J. Stawicki.   U.S. Patent 4,543,155. 1985.
6. O. Simanson, L. A. Linsdtrom and A. Marklund. Tappi Journal. 1987. August '73.
7. The Chemical Engineer. 16.1.92. 511 p.12.
8. R. W. Watson. Eur-Pat-Appl. 0218411.

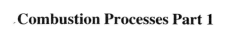

Combustion Processes Part 1

Oxygen Transfer in Fluidised Combustion: Basic Science and Its Application in Industry

J. F. Davidson

DEPARTMENT OF CHEMICAL ENGINEERING, UNIVERSITY OF
CAMBRIDGE, PEMBROKE STREET, CAMBRIDGE CB2 3RA, UK

1   SUMMARY

Mechanisms governing combustion of carbon in air-fluidized inert particles at 700° C to 950° C are reviewed. The resulting theory predicts the frequency response of a large pressurised combustor.

With propane in the air supply – to simulate coal volatiles – the carbon may burn faster as bed temperature is decreased, a surprising result.

2   FLUIDIZED BED COAL COMBUSTOR: OPERATING CONDITIONS

Figure 1 depicts the conditions in a bubbling fluidized coal combustor. A bed of inert particles, typically sand, is fluidized by air, much of it passing through as bubbles which cause violent agitation of the particles. The bed temperature is 800 to 950° C and the violent agitation of the particles causes rapid heat transfer to immersed tubes not shown in Figure 1: within the tubes there is boiling water to remove the heat of combustion. Coal is fed continuously, either blown in at the bottom or spread over the top surface of the bed. The carbon inventory is usually quite small: typically 0.5% of the particles will be carbon and the question arises, why is the carbon inventory so small? The answer is not straightforward but is helpful in showing the factors that govern the operation of a bubbling fluidized bed combustor. It is helpful to delineate the steps of a designer working from first principles, as follows.

(1) The primary bed particles are often 0.5 to 1 mm diameter. Smaller particles would be subject to elutriation. Larger particles would give a low heat transfer coefficient to the immersed tubes and hence too much of the bed would be occupied by heat transfer surface: many experiments[1] show that the heat transfer coefficient falls off as particle diameter increases, due to the gas film between the particles and the immersed surface: this film thickness is roughly proportional to particle diameter which must therefore not be too large.

(2) Given that the primary bed particles are about 1mm diameter, simple calculations show that for an air-fluidized bed at 900° C, the free-falling velocity U_t of a sand particle is about 9 m/s for air at 1 bara and about 4 m/s for air at 10 bara. Allowing for the fact that there must be a range of particle sizes and that for these

relatively large particles U_t is of order $10U_{mf}$, where U_{mf} is the incipient fluidizing velocity, then the fluidizing velocity of the operating bed has to be about 1 m/s. This gives vigorous fluidization, $U > U_{mf}$, but minimal elutriation, $U < U_t$.

(3) The choice of fluidizing velocity, about 1 m/s, determines the bed heat release. Coal is comprised of carbon and hydrogen whose calorific values are widely different: about 34 MJ/kg for carbon; about 120 MJ/kg for hydrogen (net). For stoichiometric combustion, the heat released per kg of oxygen is similar for the two elements, about 13 MJ when the fuel is carbon and about 15 MJ when the fuel is hydrogen. Hence the heat release per kg of oxygen is almost independent of the fuel. Taking a value of 13 MJ/kg oxygen, then for a bed fluidized by air at a superficial velocity of 1 m/s (1 bar, 900° C), the heat release per square metre of bed is about 0.8 MW; the corresponding figure for a bed at 10 bar is 8 MW on account of the fact that the high pressure bed receives 10 times as much air per m^2 of grate.

(4) The carbon inventory is controlled by the carbon burn-out time. Under representative conditions, see below, the burn-out time for a 1 mm carbon particle is about 200 sec. The above-mentioned air velocity of 1m/s corresponds, at 1 bar pressure, to an oxygen input of about 0.07 kg/m^{2}s, so that during the burn-out time, 200 sec, the carbon input per m^2 grate must be 0.07 x 200 x 12/32, *i.e.* about 5 kg/m^2. Since the carbon residence time is 200 sec or less, an upper limit to the carbon inventory per m^2 grate is 5 kg. With a bed depth of 1 metre, typical for a large bubbling bed, the inventory of inert material *e.g.* sand will be around 1000kg, so the carbon inventory cannot be more than about 0.5%. With a pressurised unit, the corresponding figure is 5%.

As will be shown below, the carbon inventory is important in considering the unsteady state response and therefore the bed control because the carbon in the bed represents a reservoir of energy. The carbon inventory is of obvious relevance to elutriation.

The above calculations also show the central importance of carbon burn-out time in determining bed behaviour. For this reason and because of its intrinsic interest, the carbon burn-out time has been the subject of much research.

3   BURN-OUT TIME OF CARBON PARTICLES

The burn-out time of the batch of carbon added to the air-fluidized bed can be measured in several ways as follows.

(1) Early experiments[2] were done by visual observation. The carbon particles in the bed of inerts are red-hot and burn-out can be measured by observing when the bright specks disappear at the surface of the bed. The burn-out is measured with a stop watch.

(2) A subsequent development[3] was to observe the CO_2 concentration in the off-gas. Burn-out was assumed to occur when either (a) the CO_2 concentration fell to zero or (b) when integration of the CO_2-time relation showed that 95% of the injected carbon had burned.[4]

(3) Where burn-out of carbon was measured in the presence of a combustible gas, *e.g.* propane,[5] neither method (1) nor method (2) will work. Measurement of bed temperature, T_b, was used. The heat generated by the burning carbon causes a

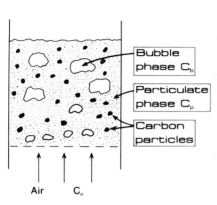

Figure 1.   Bubbling fluidized coal combustor: oxygen concentrations.

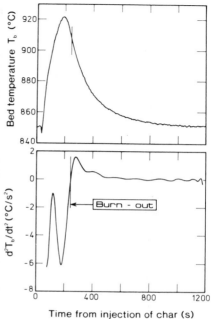

Time from injection of char (s)

Figure 2.   Combustion of char in air-fluidized sand: bed temperature and its second derivative.[5]

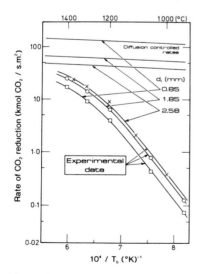

Figure 3.   Reaction of pure CO_2 with fluidized carbon.[6]

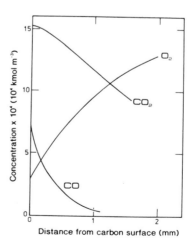

Distance from carbon surface (mm)

Figure 4.   Concentration profiles around a 2mm dia burning char particle in an air-fluidized bed.[3]

rise in bed temperature, but as the carbon burns out the relation between bed temperature and time t reverts to a cooling curve. Figure 2 shows a typical T_b-t relation, with its second derivative. The point of reversion to the cooling curve is the instant of burn-out, assumed[5] to be when $d^2T_b/dt^2 = 0$; arguably it would be better to take the slightly later time when a higher derivative of T_b is zero.

Factors affecting burn-out time

The theoretical equation[3] to predict burn-out time τ for a batch of mass m, containing mono-sized carbon particles of diameter d_i and solid density ρ_p, is

$$\tau = \frac{\rho_p d_i}{24 k_s c_o} + \frac{\rho_p d_i^2}{48 Sh\, D_g\, c_o} + \frac{m}{12 c_o A [U - (U - U_{mf})\, e^{-x}]} \qquad (1)$$

$$\begin{pmatrix}\text{Chemical}\\ \text{kinetics}\end{pmatrix} \quad \begin{pmatrix}\text{Local}\\ \text{diffusion}\end{pmatrix} \quad \begin{pmatrix}\text{Bubbles to particulate}\\ \text{phase and stoichiometry}\end{pmatrix}$$

Here k_s is the reaction rate constant between carbon and oxygen, c_o is the inlet gas phase oxygen concentration; Sh is the Sherwood number = $k_g d/D_g$ where k_g = particle mass transfer coefficient, d = carbon particle diameter, D_g = gas phase diffusion coefficient. U = superficial gas velocity at bed temperature, $U = U_{mf}$ at incipient fluidization, A = bed cross-section and X = interphase transfer factor. The three terms on the right-hand side of the Eq. (1) represent the factors controlling combustion, as follows.

(i) *Chemical rate.* For bed temperatures up to at least 900° C, the rate of combustion of carbon to CO at the carbon surface appears to be rate controlling:[3] the CO evolved appears to burn close to the carbon surface, forming CO_2 in the presence of excess oxygen and such excess is usually available during burn-out experiments. In earlier work[2] it was assumed that the reduction of CO_2 by carbon was a significant factor. But Patel,[6] who reacted pure CO_2 with carbon in a fluidized bed, found that the reaction is slow at temperatures up to 1400° C, see Figure 3: this shows that even at 1400° C the reaction of CO_2 with carbon is partly controlled by chemical rate and partly by diffusion. Therefore at 1000° C and below, as expected in a fluidized combustor, the rate of reaction of CO_2 with carbon should be negligible. This leads to the concentration profiles round the carbon particle postulated[3] in Figure 4. Note the finite concentration of CO_2 adjacent to the particle surface, consistent with the above-mentioned finding that, at the relevant temperature, CO_2 reacts only slowly with carbon. The finite concentration of oxygen adjacent to the particle is consistent with the hypothesis that the rate controlling step is the carbon/oxygen reaction adjacent to the carbon surface of rate constant k_s. The CO generated by this reaction diffuses away from the surface, reacting with incoming oxygen: the large amount of heat generated by this combustion reaction $CO + \frac{1}{2} O_2 \rightarrow CO_2$ is generated near the surface of the carbon and contributes to the burning particle heat balance in an equivocal way, see below.

(ii) *Local mass transfer around the burning particle.* Although the particles in the bubbling bed are violently agitated, the interstitial flow of gas around an individual particle, such as a burning carbon particle, is tranquil. It is easy to show that the local Reynolds number, based on gas density and viscosity, incipient fluidizing velocity and particle diameter, is of order 10. Consequently the particle Sherwood number Sh, which appears in Eq. (1), is of order 2 to 4, not much different from the value of 2 appropriate for a particle in stagnant gas. Indeed the Sherwood number could be less than 2 for a burning particle, because the inert (sand) particles, surrounding the burning particle, reduce the gas volume available for diffusion. Comparing the first and second terms in Eq. (1), the first gives a burn-out time proportional to initial particle diameter d_i whereas the second term gives $\tau \propto d_i^2$; the latter result arises because the Sherwood number is constant, so the local mass transfer coefficient k_g is inversely proportional to particle diameter. This implies that as the particle burns down, chemical reaction will control combustion rate, see below. The effect of pressure P is also of interest: c_0 the oxygen concentration is proportional to P but the gas phase diffusion coefficient D_g is inversely proportional to P, so $D_g c_0$ should be independent of P. The chemical rate term in Eq. (1) by contrast is proportional to 1/P because $c_0 \propto P$. It follows that as P is increased the chemical rate term becomes faster and at high pressure, combustion should be controlled by the diffusion term which is independent of pressure. This will be discussed below.

(iii) <u>Bubble to particulate phase transfer of oxygen</u>. The last term in Eq. (1) represents diffusion resistance governing mass-transfer of oxygen from the bubble phase to the particulate phase. The interphase transfer term also represents a stoichiometric factor: even if $X \rightarrow \infty$, the burn-out time increases with m because a larger carbon charge implies that more oxygen is needed for combustion.

4   EXPERIMENTAL RESULTS FOR BURN-OUT TIME

<u>Single particle burn-out time</u>

Figure 5 shows typical results from carbon burn-out time experiments: τ is plotted against carbon charge m; the relation is linear, with a finite intercept at m = 0. This finite intercept has been called the 'single particle burn-out time' τ_s[7]: τ_s is the burn-out time for a single carbon particle placed in the bed; assuming m = 0 in the experiment (an assumption which is not quite valid for a large carbon particle), the interphase transfer term is negligible and combustion of the single particle is governed by (a) chemical kinetics and or (b) local diffusion near the particle, the first two terms on the right-hand side of Eq. (1). The relative importance of (a) and (b) depends upon the bed temperature and the nature of the carbon. Campbell[7] plotted τ_s against d_i^2: if there was diffusion control, τ_s would be proportional to d_i^2; in fact the relation is, from Eq. (1), of the form $\tau_s = Bd_i + Ed_i^2$, B and E are constants for given conditions of bed temperature and carbon. Figure 6 shows that this theoretical relation fits the data quite well: evidently chemical rate is the controlling step for small particles. The longer burn-out times for coke as compared with char must be because the latter particles were more reactive.

<u>Effect of pressure</u>. Figure 5 shows the effect of pressure on burn-out time, quite small for coke, Figure 5(a), but very substantial for char derived from a high volatile coal, Illinois no. 5, Figure 5(b). As noted above, increased pressure increases the chemical rate, whereas the mass transfer term is independent of pressure. For the Illinois no. 5 char, chemical rate is evidently important at atmospheric pressure, but at 12 bar and above the combustion is controlled by mass transfer. Note that the data in Figure 5(b) were for a lower temperature, 850° C, conducive to chemical rate control, as compared with the results for coke Figure 5(a) at 900° C for which the chemical rate effect was less important.

<u>Effect of carbon charge m</u>. Because $c_0 \propto P$, Eq. (1) predicts that $\tau \propto m/P$; the data in Figure 5 show that this is indeed the case, the primary effect being that of stoichiometry; a larger carbon charge needs more oxygen. Likewise there is an

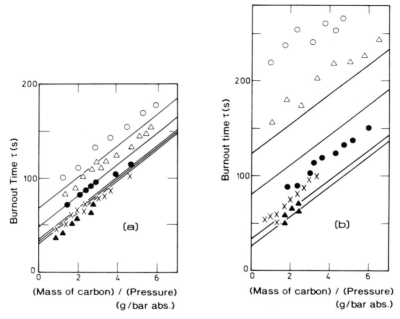

Figure 5. Burn-out times in a pressurised air-fluidized bed. The lines
are from theory.[4] (a) Coke, $T_b = 900°$ C; (b) Char, $T_b = 850°$ C.

Data point	O	Δ	●	X	▲
Pressure (bara)	1.2	2	7	12	17
(a) Coke 1-1.4mm: estimated T_c (°C)	1022	1062	1100	1108	1112
(b) Char 0.7-1mm: estimated T_c (°C) (Illinois no 5)	907	932	1010	1038	1052

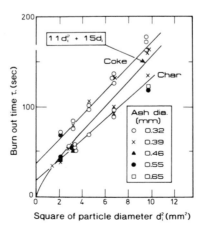

Figure 6. Burn-out time for a single
carbon particle in air-fluidized ash.

opposite effect of pressure, higher pressure bringing in more oxygen at constant superficial velocity.

Effect of bed temperature. Figure 7 shows that the effect of bed temperature on burn-out time varies widely with the type of carbon. For coke, relatively impervious carbon, increase of bed temperature causes a dramatic reduction of carbon burn-out time: evidently the combustion is chemical rate controlled at low bed temperatures and highly sensitive to temperature. By contrast, the char from Illinois no. 5 coal was highly reactive, due to its porous structure: hence diffusion control must have been dominant at all temperatures and burn-out time is therefore almost independent of bed temperature.

5   CARBON PARTICLE TEMPERATURE; COMBUSTION OF SMALL PARTICLES

The carbon temperature T_c is governed by a heat balance, the temperature T_c reaching a value such that the rate of heat generation, by carbon oxidation, equals the rate of heat loss by convection and radiation, giving the following equation.[4]

$$\Delta H c_p/(1/k_s d + 1/Sh\ D_g) \quad = \quad Nu\ k(T_c - T_b) \quad + \quad \varepsilon\sigma d(T_c^4 - T_b^4) \qquad (2)$$

$$\begin{pmatrix}\text{Heat generation from}\\ \text{carbon combustion}\end{pmatrix} \qquad \begin{pmatrix}\text{Heat loss}\\ \text{by convection}\end{pmatrix} \qquad \begin{pmatrix}\text{Heat loss}\\ \text{by radiation}\end{pmatrix}$$

Here ΔH is the heat generated at the surface of the carbon by combustion and c_p is the particulate phase concentration of oxygen. Nu is the local Nusselt number = hd/k where h = local heat transfer coefficient and k = thermal conductivity of the particulate phase; ε = carbon emissivity and σ = Stefan-Boltzmann constant.

This type of heat balance was applied by Ross *et al* [8]: results are shown in Figure 8; data were obtained by colour matching between (i) carbon particles burning in a fluidized bed and (ii) a heated tungsten filament at known temperature. The linear relation between $T_c - T_b$ and oxygen concentration c_p is expected. Note that the temperature T_c is at most 140° C above bed temperature, confirming 'cool combustion'; the carbon temperature, 1000-1100° C, is much below what is expected for pulverised fuel combustors.

The explanation for the 'cool combustion' can be discerned from Eq. (2). For an isolated carbon particle – as in a pulverised fuel burner – Nu and Sh are of the same order of magnitude. For the fluidized combustor, Nu is of order 7 whereas Sh is of order 2-3[4]: the difference arises because of the inert particles which assist heat transfer but impede mass transfer. Moreover, the effective thermal conductivity for heat transfer from a carbon particle is that of the gas/solid mixture, somewhat higher than that of the gas. The upshot is that heat transfer from a carbon particle is altogether more effective than mass transfer and hence the small temperature difference $T_c - T_b$.

Figure 9 shows how estimated carbon temperature, using Eq. (2), varies with particle diameter and with operating pressure. The form of the curves arises from the interplay of the terms in Eq. (2) and the salient features of the curves can be explained as follows.

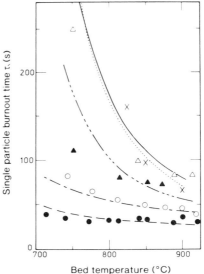

Figure 7. Effect of bed temperature on the burn-out time of a single carbon particle at atmospheric pressure.[4]

Carbon:	Coke	Merthyr Vale	Illinois 5	Bedwas	Tymar
	×	O	●	▲	△
Theory[4]	.....	— — —	—	—·—·—	— —

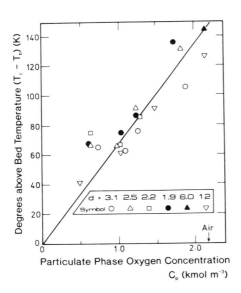

Figure 8. Measurements of (Carbon temperature) - (Bed temperature) against oxygen concentration.[8]

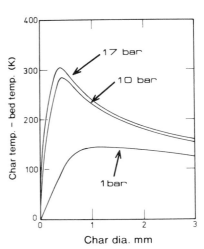

Figure 9. Theoretical (Carbon temperature) - (Bed temperature) against particle diameter.[9] Effect of pressure.

(i)   The increased particle temperature with high pressure – also noted in the caption to Figure 5 – arises because of the chemical rate term in Eq. (2) which increases with pressure because of the higher oxygen pressure.

(ii)   For large particle diameters, the radiant heat transfer beam becomes dominant, being proportional to particle diameter d: all other terms diminish or stay constant as particle diameter increases, so the carbon particle temperature tends towards bed temperature T_b for very large d.

(iii)   For very small d, $T_c \rightarrow T_b$, so there must be a maximum in the relationship between T_c - T_b and d. For very small carbon particles, the chemical rate term becomes small in comparison with the convective term so the carbon temperature becomes that of the bed. This fact is important for designers of fluidized combustors: the fact that small carbon particles are cooler than big ones means that the small carbon particles burn slowly and are therefore liable to be elutriated from the bed.

(iv)   Fast fluidized bed combustors may be expected to give higher carbon particle temperatures. For them the inert particles are relatively remote from each burning carbon particle so it is to be expected that $Sh \simeq Nu$ and T_c - T_b is likely to be bigger than for the bubbling bed. There are two consequences namely (1) carbon combustion efficiency should be higher and (2) NO_x generation may also be greater. These are speculations which need to be checked against measurements.

6   CONTINUOUS CARBON FEED: PARTICLE SIZE DISTRIBUTION IN THE BED

A combustor fed continuously with fuel will exhibit a carbon particle size distribution in the bed. It is instructive to consider size distributions generated in a bed fed continuously with mono-sized carbon: the result depends on whether the combustion is controlled by chemical kinetics or diffusion. The analysis ignores particle breakage, an important factor with large coal. The analysis assumes that all particles experience the same environment irrespective of size, a plausible assumption apart from small particles which get elutriated.

Combustion controlled by chemical kinetics

In this case the rate of change of carbon particle radius $dr/dt = -C = $ constant. The number of particles in the size range (r, r + dr) is Qdr, which defines Q. For a steady state it follows that $(Q + dQ)\Delta r = Q\delta r$: particles from the size interval (r + dr, r + dr + Δr) enter the size range (r, r + dr) during time dt; particles from the size interval (r, r + δr) leave the size range (r, r + dr) during time dt. Now $\Delta r = Cdt$ and $\delta r = Cdt$, so $dQ = 0$ and

$$Q = \text{constant} \qquad (3)$$

The conclusion is that when combustion is controlled by chemical kinetics, the number in size range dr is independent of r.

Combustion controlled by diffusion

When the rate of oxygen transfer to each carbon particle controls its combustion rate, a material balance on a carbon particle must give consistency between the rate of change of particle volume and the rate of diffusion of oxygen to the particle, so that

$$\rho_p \frac{d}{dt} \left(\frac{4}{3} \pi \, r^3 \right) = -k_g \, 4\pi \, r^2 \, c_p \tag{4}$$

As noted above, the Sherwood number, $Sh = k_g 2r/D_g$, is of order 2. It follows that $k_g \propto 1/r$ and hence from Eq. (4),

$$dr \, / \, dt = (\text{constant}) \, / \, r \tag{5}$$

Eq. (5) in combination with the function Q, defined above, gives

$$Q = (\text{constant}) \, r \tag{6}$$

Comparing Eqs. (3) and (6), the size distribution is very different depending on whether the carbon combustion is controlled by chemical kinetics or by diffusion, the latter giving a smaller proportion of fine particles because they burn rapidly with diffusion control. Campbell[7] demonstrated that results for coke particles followed Eq. (6) see Figure 10, *i.e.* diffusion control was exhibited; but the method is difficult to apply for combustors in general because (a) the feed usually contains a wide size range (b) the analysis is difficult: it is necessary to sample the bed, quenching with inert gas to preserve the in-bed size distribution.

7   UNSTEADY STATE BEHAVIOUR: CONTROL OF FLUIDIZED COMBUSTORS

Gray[10] gives an analysis of unsteady state behaviour of fluidized bed combustors, compared with data for laboratory, pilot and industrial units. He considered a bed with a coal or carbon feed of the form $N + n \exp(i\omega t)$ where N and n are constants, and ω is the circular frequency. He showed that the combustor can be modelled by three first-order terms, so that the amplitude and phase lag for temperature oscillations are given by

$$A_t = \sqrt{[(1 + \omega^2 \tau'^2)(1 + \omega^2 \tau_b^2)(1 + \omega \tau_t^2)]} \tag{7}$$

$$\varphi_t = \tan^{-1} \omega \tau' + \tan^{-1} \omega \tau_b + \tan^{-1} \omega \tau_t \tag{8}$$

The time constants τ', τ_b and τ_t are as follows.

(1)   The time constant τ' is a characteristic combustion time = (fuel inventory)/(fuel feed rate). For a mono-sized feed of burn-out time τ, Gray[10] used the results of van der Post *et al* [11] to show that τ' and τ are related according to the mechanism controlling combustion so that (a) for diffusion control, $\tau' = 2\tau/5$, (b) for chemical rate control, $\tau' = \tau/4$ and (c) for interphase mass transfer control, $\tau' = \tau/2$.

(2)   The thermal time constant of the bed τ_b represents the time for a fuel-free unit to respond to temperature changes, so τ_b = (Heat capacity of bed) / (Rate of heat loss): the rate of heat loss is governed by the factors that remove heat from the bed, following a temperature change, namely (i) the product of heat transfer coefficient × surface area for the cooling surfaces, tubes and wall and (ii) the product of mass flow rate × specific heat for the fluidizing gas.

(3)   The time constant τ_t is the response time of the in-bed thermocouple.

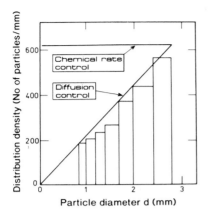

Figure 10. Continuous feed of mono-sized carbon: resulting size distribution in the bed.[7]

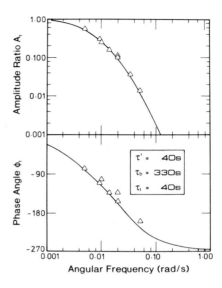

Figure 11. Bode diagram for pressurised combustor. Data from IEA unit, compared with Equations (7) and (8).[10]

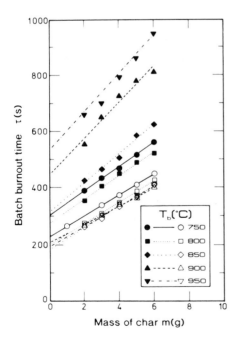

Figure 12. Char burn-out times in sand for (a) air-fluidized bed O □ ◊ Δ ∇; (b) bed fluidized with air/2.5% propane ● ■ ◆ ▲ ▼.[5]

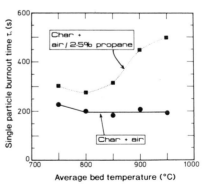

Figure 13. Single particle char burn-out time against temperature for: (a) air-fluidized sand (b) sand fluidized by air/2.5% propane.[5]

Frequency response for a pressurised combustor

Gray[10] reports a comparison between the above theory and results from the 2 m × 2 m cross-section fluidized combustor at Grimethorpe, Yorks, UK of the International Energy Agency. The coal feed rate was varied sinusoidally with a bed depth of 4.5 m, pressure 12 bar and fluidizing velocity 2.5 m/s. Figure 11 shows the results in comparison with Eqs. (7) and (8). The values of $\tau_b = 330$s and $\tau_t =$ 40s were deduced from other experiments. The coal feed contained particles of size 0 to 6 mm with a median size (wt) of about 1 mm. The value of $\tau' = 40$s chosen by Gray[10] may be compared with the data in Figure 5 for burn-out of coke or char in a similar size range: the burn-out time with a substantial charge is of order $\tau = 100$s. As noted above, $0.25 \leq \tau'/\tau \leq 0.5$ depending upon the mechanism controlling combustion, so $\tau' = 40$s is consistent with the data in Figure 5. With these time constants, the theory is in remarkably good agreement with the data in Figure 11, bearing in mind that the results are from an industrial scale unit.

8   THE EFFECT OF VOLATILES ON THE COMBUSTION OF CHAR

When bituminous coal enters a fluidized bed, the volatiles are released rapidly – within 10 to 15s for a 1mm particle – and compete with the residual char for the available oxygen. Thus the volatiles must increase the char burn-out time. Experiments to clarify the effects were done by fluidizing hot sand with a mixture of excess air and propane; depending on the bed temperature, the propane burned in or above the bed.[5] Carbon particles were then added and their burn-out time deduced from bed temperature measurements as described above, burn-out being inferred when $d^2T_b/dt^2 = 0$. Whether the propane is representative of volatiles from coal is a moot point: the average molecular weight of coal volatiles may be of the same order as propane, but coal volatiles have compositions ranging from tars to light hydrocarbons.[12] However, the propane experiment does give well-defined conditions and may have bearing on the coal volatiles/char combustion question.

Figure 12 shows typical results for char burn-out time with and without propane in the air supply to the bed. The propane increases the burn-out time by a factor of 2 or more in extreme cases. However, at the lower temperatures of order 750° C, the increase of burn-out time is not so great as at high temperatures 900 to 950° C. This must be because the propane burns more effectively in the bed at high temperatures. There is evidence[13] that at temperatures below 800 to 850° C, the propane does not burn in the particulate phase of a fluidized bed. There seems no doubt that over-bed burning is more likely at low bed temperatures. At high temperatures, 900° and above, there is evidence that the propane burns quite close to the air distributor, thus lowering the oxygen concentration in the bed and increasing the burn-out time of carbon particles in the bed.

Figure 13 shows the effect of bed temperature on char burn-out time with and without propane in the air supply. With propane, there is the rather striking result that the burn-out time actually increases with rise of bed temperature *i.e.* char combustion is slower at high temperature. This apparent departure from the usual effect of temperature is because the propane burns more readily at high temperature, thus reducing the oxygen concentration around each char particle. Hence it may not be beneficial to raise the temperature of a coal combustor with fuel of high volatile content. At a low bed temperature, much of the volatile content will burn above the bed and suitable means of removing the heat must be provided. But the carbon combustion efficiency – with regard to in-bed combustion – may be better than with higher bed temperatures, which would obviate the need to collect and burn the fine carbon elutriated from the bed.

NOTATION

A	bed area	r	particle radius
A_t	amplitude lag	Sh	Sherwood number
B	constant	T_b	bed temperature
C	constant	T_c	carbon temperature
c_o	inlet oxygen concentration	t	time
c_b	oxygen concentration in bubble	U	superficial fluidizing velocity
c_p	particulate phase oxygen concentration	U_{mf}	U at incipient fluidization
d	particle diameter	X	cross-flow factor
D_g	gas phase diffusion coefficient	ΔH	heat of carbon combustion
d_i	initial particle diameter	ε	particle emissivity
E	constant	ρ_p	carbon density
k	thermal conductivity, gas or particulate phase	σ	Stefan-Boltzmann constant
k_g	mass transfer coefficient	τ	carbon burn-out time
k_s	surface rate constant	τ'	combustion time constant
m	mass of carbon charged	τ_b	bed time constant
N	carbon feed rate	τ_s	single particle burn-out time
Nu	Nusselt number	τ_t	thermocouple time constant
n	carbon feed fluctuation	ϕ_t	thermal phase lag
P	pressure	ω	circular frequency
Q	(particle number)/dr		

REFERENCES

1. J.S.M. Botterill, 'Fluid-Bed Heat Transfer', Academic Press, 1975, p 231.
2. M.M. Avedesian and J.F. Davidson, Trans.Instn.Chem.Engrs., 1973, 51, 121.
3. I.B. Ross and J.F. Davidson, Trans.Instn.Chem.Engrs., 1982, 60, 108.
4. E. Turnbull, E.R. Kossakowski, J.F. Davidson, R.B. Hopes, H.W. Blackshaw, and P.T.Y. Goodyer, Chem.Eng.Res.Des., 1984, 62, 224.
5. R.P. Hesketh and J.F. Davidson, Chem.Eng.Sci., 1991, 46, 3101.
6. M.S. Patel, 'Fluidized Bed Gasification and Combustion', PhD dissertation, Cambridge (1979), Fig. 5.4.
7. E.K. Campbell and J.F. Davidson, Inst.Fuel.Symp. Ser., 1975, no 1, paper A2
8. I.B. Ross, M.S. Patel and J.F. Davidson, Trans.Instn.Chem.Engrs., 1981, 59, 83.
9. J.F. Davidson, Inst.Energy Symp. Ser., 1984, no 3, vol 2, REV/1/1.
10. D.T. Gray, 'The Control of Fluidised Combustors', PhD dissertation, Cambridge (1986), pp 32, 61, Fig 45.
11. A.J. van der Post, O.H. Bosgra and G. Boelens, Inst. Energy Symp. Ser. , 1980, no 4, paper IV-3.
12. J.F. Stubington and S.W. Chan, Trans.Instn.Chem.Engrs., 1990, 68A, 195.
13. R.P. Hesketh and J.F. Davidson, Combustion and Flame, 1991, 85, 449.

Influence of Oxygen on Ignition and Combustion of Single Coal and Char Particles

D. Schwartz, R. Gadiou, and G. Prado

LABORATOIRE GESTION DES RISQUES ET ENVIRONNEMENT, RUE DE
CHEMNITZ, 68067 MULHOUSE CÉDEX, FRANCE

1. INTRODUCTION

During the last decade, several different techniques
were developed for coal combustion :
- pulverised coal flame, mainly used for electricity
generation
- in steel manufacturing, increasing quantities of
coal are injected in blast furnaces.
- fluidized bed combustion has been developed as a
means for better pollutant control and is used in
many processes (coal combustion, gasification to
produce high BTU gases,...)

These different ways of burning coal use very
different conditions of temperature and oxygen contents of
the combustion atmosphere. The steel manufacturers, for
example, try to use high oxygenated combustion
atmospheres. Among the different processes involved in
coal combustion, many of them are influenced by the oxygen
content of the atmosphere, for example :
- the ignition of the fuel particle.
- the chemical reaction of coal volatiles in the
gas phase, including formation of soot.
- the formation of char during the
devolatilization which may be influenced by the
energy feed back of volatile matter combustion.
- the heterogeneous reaction between char and
oxidizing gas species.

The influences of these effects can be observed in
burners and combustors on flame structure and temperature
and on pollutant formation during coal combustion.

In order to study the effects of oxygen on coal
combustion, we have developed two different reactors which
allow to work on a single particle : a drop tube furnace
is used for pulverised coal combustion, with coal diameter
ranging from 80 to 350 μm and a laser reactor for samples
having diameters between 0.5 and 3 mm. The information
obtained with these reactors is based mainly on two types

of measurements : combustion times and temperatures, achieved by analyzing the light emitted by the burning sample. Pollutant analysis is performed by batch experiments or with on-line analysers. These experimental values are compared with predictive models developed in the laboratory.

2.   PULVERISED COAL COMBUSTION

Large amounts of coal are burnt in pulverised coal flames for electricity generation or blast furnaces firing. Modelization of the combustion of micrometer sized coal particles is of first importance to correctly predict industrial operations.

In order to study the pulverised coal combustion, the experimental features which are generally used are : entrained flow [1] [2] [3] , fixed beds [4] [5] , fluidized beds [6] [7] , thermobalance[6] [8] [9] . In these types of reactors, interactions between particle reactivity and diffusion processes in a two phase medium are important and very difficult to modelize. Extensive reviews of coal studies have been published by Essenhigh [10], Smoot [11] and Laurendeau [12] for example.

Single particle experiments give results which are easier to analyse as they can be correlated to predictive models. We have developed a new drop tube furnace which can be used to study the combustion of single particles. Measurements are done all along the combustion by analyzing the emitted light with an optical pyrometer.

2.1. THE DROP TUBE FURNACE

The experimental furnace is described in figure 1. The reactor is a vertical cylindrical alumino - silica tube with a 4.5 cm internal diameter and a 140 cm length. The heating is performed by six lanthanum chromite bars, providing a maximal power of 12 kW which allow a maximal wall temperature of 1700 K.

Coal particles are individually injected axially at the top of the reactor by the mean of a distributing plate and they fall through a water cooled probe in the preheated gas stream. With particle sizes of nearly one hundred microns, heating rate is between 10^5 and 10^6 K/s. The combustion gas composition is controlled by a mass flow meter providing a highly stabilized oxygen content of the gas. When the coal particle enters the hot gas stream, it crosses a laser beam and its partial extinction is detected by a photo cell. This generates a trigger signal, used to define the zero time to start the acquisition by the micro computer.

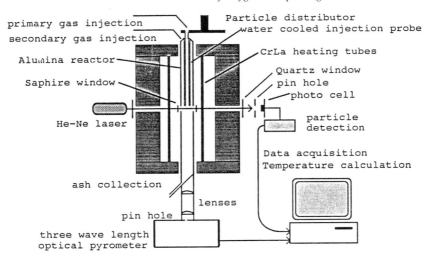

primary gas injection — Particle distributor
secondary gas injection — water cooled injection probe
Alumina reactor — CrLa heating tubes
Saphire window — Quartz window
pin hole
photo cell
He-Ne laser — particle detection
Data acquisition
Temperature calculation
ash collection
lenses
pin hole
three wave length
optical pyrometer

Figure 1 : The drop tube furnace

A three wavelength pyrometer mounted axially at the bottom of the reactor analyses the light emitted by the burning particle. This allows to measure three characteristic periods of the combustion : ignition delay, volatile matter flame duration and heterogeneous combustion duration. The particle surface temperature is then calculated from the ratio of particle emittance assuming a grey body behaviour. The three wavelengths are selected by three narrow band path interference filters centred at 600, 694 and 746 nm. Typical curves of monochromatic light emission and computed temperature are presented in figure 2a and 2b.

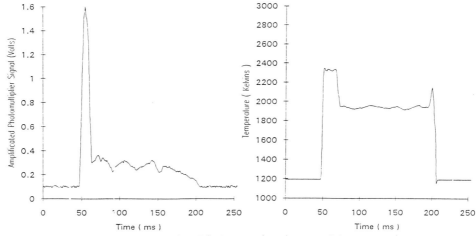

Fig 2 : a) Monochromatic light emission , b) Particle surface temperature

In order to take into account heterogeneity in particle size, shape or composition, nearly a hundred experiments are done for each experimental condition of furnace temperature, sample composition or size and gas composition. We can obtain precise experimental values by statistical analysis.

2.2. EXPERIMENTAL RESULTS

2.2.1. Ignition delay

Measurements of ignition times have been done for four coal samples, their proximate and elemental analysis are given in table 1. Three particles sizes were selected : 80 - 125 μm, 125 - 160 μm and 160 - 200 μm. Experiments were conducted with furnace temperatures ranging from 1073K to 1473 K, and O_2 mole fractions of the gas from 0 to 100%.

coal	ash %mf	V.M.%daf	Elemental Analysis (%daf)				
			C	H	O	N	S
Gardanne	24.9	37.4	71.0	4.7	18.9	1.7	3.9
Australian	12.8	24.3	77.5	5.1	1.8	14.6	1.0
Colombian	8.9	38.3	76.7	5.8	1.7	14.9	0.9
South African	15.5	25.3	87.0	4.7	5.2	0.6	2.5

Table 1: Analysis of the four samples

Measured ignition times are shown in figure 3. We can see that the oxygen content of the combustion gas has little influence as compared to the furnace temperature.

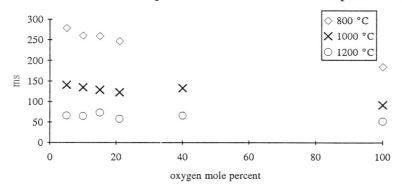

Figure 3 : Measured ignition delay

In most experiments, the influence of particle diameter is not very important. This is a consequence of the difference in drop speed limit between the different particle sizes. A larger coal particle falls with a higher speed and enters more rapidly in the hot gas stream and a small particle stays longer in the heating zone of the entraining gas flow. This influence balances the difference in mass/volume ratio between small and large particles, so that ignition times are nearly constant. This fact had already been observed in our laboratory with other coals [13].

2.2.2.   Volatile matter flame

The influence of oxygen on volatile matter flame was studied with the four coals already described. For particle sizes lower than 100 microns no flame was observed for all oxygen concentrations and temperatures studied. Devolatilization occurs during heterogeneous combustion, but the volatile matter flow isn't sufficient to stabilize a diffusion flame around the particle. Similarly, at a furnace temperature of 800°C, there is no flame for all the particle sizes studied. If the sample inflammation occurs, it must be related to an heterogeneous ignition process. This fact can also be observed at 1000°C, but only for the smaller particles.

Figure 4 presents measured flame duration for a Gardanne coal as a function of oxygen content of the combustion gas. Below 10% oxygen, there is a very important dispersion of the values; this fact is related to the difficulty to ignite the evolving volatile matter. Above 10% O_2, we can observe an exponential decrease of flame duration.

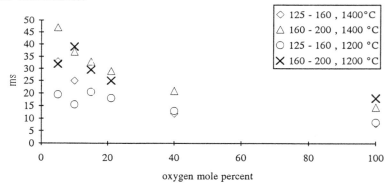

oxygen mole percent

Figure 4 : Measured flame duration

2.2.3.   Heterogeneous combustion

A first set of measurements was made with the four coals previously described. The experimental combustion times are given in figure 5 for Gardanne coal (particle diameter 100 - 125 μm). We can see that there is an exponential decrease when the oxygen content of the gas increases and that furnace temperature has very little influence.

This is an indication of diffusional control for the heterogeneous reaction rate. Similar behaviours were observed for the four coals. Combustion temperatures are given in figure 6 for Gardanne coal. The influence of oxygen on reaction rate is well illustrated by the increase in particle temperature.

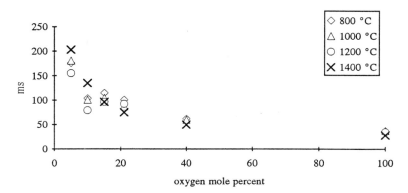

<u>Figure 5</u> : Combustion time. (Gardanne 125 - 160 μm)

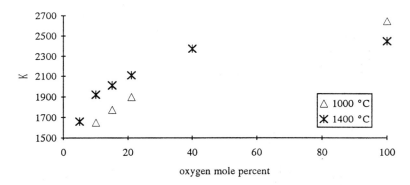

<u>Figure 6</u> : Combustion temperature (Gardanne 125-160 μm)

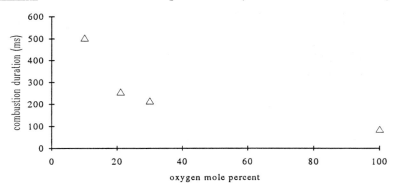

<u>Figure 7</u> : Combustion time (Freyming 200 -250 μm)

A second set of experiments were done with another coal from Freyming [14] (ash : 3% mf; M.V. : 33.6% daf,

elemental analysis (% daf) : C : 82.9 ; H : 5.5 ; O : 9.3
; N : 1.15 ; S : 0.7). Combustion times at 1273 K for 200
- 250 μm particles diameter is given in figure 7. Again
furnace temperature has no influence. The combustion rate
is also controlled by the diffusion of oxygen in the
boundary layer.

2.3. MODELLING OF THE COMBUSTION OF A CHAR PARTICLE

In order to analyse these experimental results, two
models were developed at the laboratory; they are both
based on the unreacted Shrinking Core Model. The first one
was used with the four coals named Gardanne, South
African, Colombian and Australian and the second was
created to correlate the experimental data of Freyming
coal. The main difference was the observation that with
Freyming coal hollow spherical cenospheres are formed
during devolatilization.

2.3.1.   Model 1 [15]

The coal particle was supposed spherical during the
devolatilization. An important fraction of the coal is
evolved from the particle and this leads to the
development of a new porosity in the sample. The
heterogeneous combustion between coal and oxygen must
account for three possible resistances as shown in fig 8:
- the diffusion of oxygen in the boundary layer of the
particle.
- the diffusion of oxygen in the ash layer.
- the kinetics of carbon - oxygen reaction.

The reaction rate with these assumptions is :

$$Rm = 4.\pi.Rp^2.\left[\frac{1}{Kd} + \frac{Rp(Rp - Rn)}{Rn.\Delta.D_e'} + \frac{Rp^2}{Rn^2.Kf}\right]^{-1}.C_0$$

Rm : global reaction rate (g/s)
Rp : particle diameter (m)
Rn : carbon core diameter (m)
Kf : intrinsic carbon - oxygen kinetics (g/m^2.s)
C_0 : bulk oxygen concentration (g/m^3)
D   : stoichiometry
D'_e: effective diffusion coefficient in ash layer (m^2/s)
Kd : diffusion coefficient in the boundary layer (m^2/s)

The computed combustion durations with this model are
given in fig 9. The agreement with experimental values is
good , except for very low oxygen mole fraction. This can
be the result of an under-estimation of the experimental
combustion duration, as the combustion temperature is low
in this case, and it is very difficult for the pyrometer
to detect the burning char in the hot furnace.

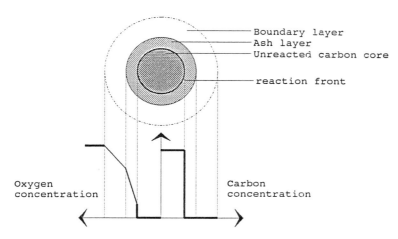

Figure 8 : Coal combustion model

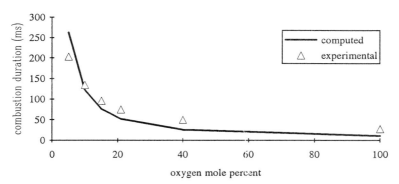

oxygen mole percent

<u>Figure 9</u> :Heterogeneous combustion durations (125-160 μm).

2.3.2.   Model 2

The main assumption of this model is described in fig 10; the devolatilization results in the formation of a cavity in the center of the particle. The char can be assumed to be a non-porous carbon cenosphere. According to microscopic observation and size measurements, we observed no swelling of the char.

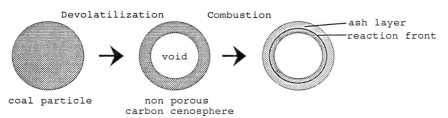

<u>Figure 10</u> : Model 2 assumptions.

The comparison between experimental and computed values is given in figure 11. The agreement is good.

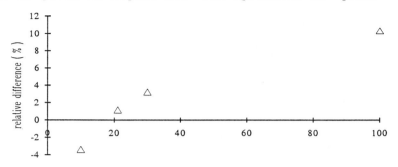

oxygen mole percent

<u>Figure 11</u> : Comparison between calculated and experimental combustion duration.

3. MILLIMETRE SIZED COAL COMBUSTION

During the last ten years, an important development of the fluidized bed combustion of coal was observed. Some of the main advantages of this technique are the use of higher sized coals, a better heat transfer in the bed and an easier elimination of pollutants generated by the combustion. In this type of combustor, the coal particles are very rapidly heated so that thermal gradients are observed in the grain and mass transport phenomena can not be neglected.

Our laboratory has developed a new laser reactor in order to study the combustion and the pyrolysis of coals particles having diameter between 0.5 and 3 millimetres. The objectives in developing this experimental feature were to obtain high heating rates of single particles, and to perform pollutant analysis.

3.1. LASER REACTOR

The experimental set up is shown in figure 12 [16]. The heating of the sample is performed by two 25 watts CO_2 lasers. The beam diameter is 3 mm; their power can be adjusted by using a pulsed control signal. The lasers are water cooled in order to get constant power. Electromagnetic shutters are used to control the heating of the particle. The I.R. beams (10.6 μm) are lined up with two 0.5 mW helium neon lasers using beam selectors. The laser beams are then focused by ZnSe lenses and are directed to the opposite hedges of the reactor. This technique allows an homogeneous heating of the sample. The beams enter in the reactor through ZnSe windows. The combustion atmosphere can be controlled by a gas mixer based on mass flow rate controllers.

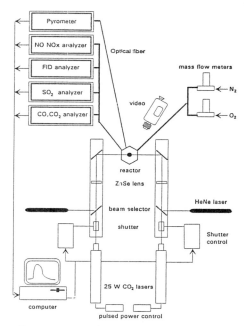

<u>Figure 12</u> : Laser reactor

During the combustion of the sample, the light emission of the coal particle can be analysed by a three wave length pyrometer through an optical fiber. The monochromatic emissions are then measured by photo multipliers through interference filters. The wavelengths used are 450 , 600 and 800 nm. Light emission can be analysed as described above with the drop tube furnace to give combustion characteristics. By the use of three colours pyrometry, we can then calculate the particle surface temperature during the whole combustion. The calibration of the pyrometer is achieved with a calibrated tungsten ribbon lamp.

Pollutant emissions can be measured by two means. In batch experiments, we can do gas sampling in the reactor through a septum in order to analyse the pollutant by gas chromatography or mass spectrometry coupled to gas chromatography. On line measurement can be done with different analysers : two way chemi-luminescent analyser for NO and NOx emission, UV fluorescence for SO_2,two way NDIR analyser for CO and CO_2 and ionisation flame analyser for unburned hydrocarbons. The whole control of the experimental features and the data acquisition and treatment is performed by a micro computer. Results on pollutant emissions are not reported in this paper, because of lack of space.

Four coal samples were used in this study, three French coals : Freyming, Vernejoul and Gardanne, and a

South African one. Table 2 sums up their proximate and
ultimate analysis. The particle diameters were ranging
from 2.5 to 3.15 mm and the laser power was 2 times 20
watts, providing a mean radiative heat flux of 570 watts /
cm^2.

			Elemental Analysis (% daf)				
sample	ash %mf	V.M.%daf	C	H	O	S	N
Freyming	3	33.6	82.9	5.5	9.3	0.7	1.15
Vernejoul	6.1	36.5	78.6	5.2	13.9	1.2	1.1
Gardanne	24.7	58.3	67.8	4.6	21.1	5.6	0.9
South African	19.6	33.0	76.7	4.3	16.6	0.4	2.0

Table 2: Analysis of the four coals

3.2. EXPERIMENTAL RESULTS

3.2.1.   Ignition delays

Ignitions delays have been measured for the four coal
samples with oxygen mole fractions ranging from 0 to 100%.
Results are summed up in fig 13. For the lower
concentrations of oxygen, ignition is very difficult so
that experimental values are very dispersed. Between 15
and 40% O_2 in combustion gas, we can observe that ignition
times are quite constant at a value of 4.5 seconds. Above
40 % oxygen, there is an important decrease for all coals
to values near 1 second or less. As observed for smaller
particle sizes, Freyming coal presents a behaviour
different from the three others samples.

In order to analyse these results we developed a
computer model of the heating of an inert sphere in the
laser reactor. The computed temperature profiles are
showed in fig 14.

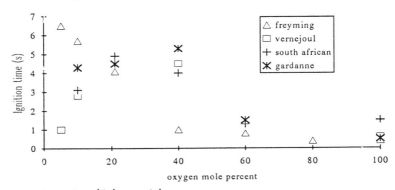

Figure 13 : Ignitions times

We can see that 4 seconds correspond to a very high
temperature of the surface of the particle. These computed
values show that 4 seconds of ignition time is

inconsistent with the hypothesis of an heterogeneous ignition.

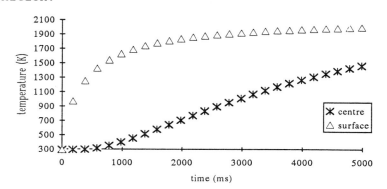

<u>Figure 14</u>: Temperature profiles (2*20 watts)

With oxygen concentration equal or less than 40% vol., the rate of the heterogeneous reaction between coal and oxygen is not sufficient to ignite the particle and to stabilize a high temperature steady state of combustion. The consequence is that ignition can only occur when substantial devolatilization has occurred and, because of the thermal gradient in the coal particle, a few seconds are needed to achieve it. As can be seen on figure 4, this corresponds to the time which is necessary to heat the inside of the coal particle at a temperature sufficient to begin devolatilization. For higher concentrations of oxidant, the kinetics of the carbon - oxygen reaction are sufficient to rapidly ignite the particle before important devolatilization is achieved.

3.2.2.   Volatile matter flame

Flame durations are shown in fig 15 for the four coals of this study. Below 10% oxygen, the inflammation of volatile matter is very difficult because the combustion atmosphere of the sample is cold.

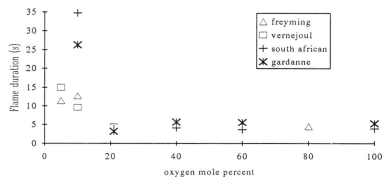

<u>Figure 15</u> : Measured flame durations

Above 15% oxygen, the ignition of the volatile matter flow is easier and the oxidant is sufficient to burn the emitted gas fuel species. In these conditions, the duration of the flame is directly related to the volatile matter flow escaping from the coal, so that this combustion characteristic still remains constant between 21 and 100% O_2. The important parameter is the volatile matter contents of the coal sample as can be seen on table 3.

Coal	South African	Freyming	Vernejoul	Gardanne
flame duration (s)	3.87	4.26	4.67	4.95
M. V. (%daf)	33.0	33.6	36.5	58.5

Table 3 : Comparison between mean flame duration (21 to 100 % O2) and volatile content of the parent coal

3.2.3.   Heterogeneous combustion

Heterogeneous combustion duration and temperatures are shown in fig 16a and 16b . We can observe an exponential decrease of combustion duration and an increase of temperature when the oxidant fraction of the gas increases. For a given oxidant content of the gas mixture heterogeneous combustion duration is directly related to the fixed carbon present in the particle after the devolatilization. Heterogeneous combustion temperatures increase constantly with the oxygen content of the gas. For 5 and 10% O_2, measured values are dispersed because of the difficulty to analyse low light emission.

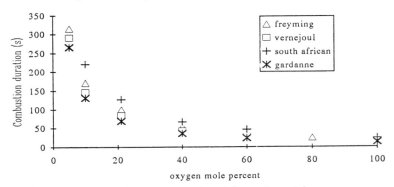

Figure 16a: Heterogeneous combustion durations

Solid particle combustion can be controlled by two processes [17]. The first one is the diffusion of the oxidant in the boundary layer of the coal particle. The heterogeneous reaction rate is inversely proportional to the sample diameter. The second one is the intrinsic kinetics of the heterogeneous solid -- gas reaction and the reaction rate is proportional to diameter. For combustion temperatures higher than 1200 K, the reaction rate is controlled by a boundary layer diffusional process [18] [19].

In this case, heterogeneous combustion duration must be related to the inverted oxygen concentration of the combustion gas [20].

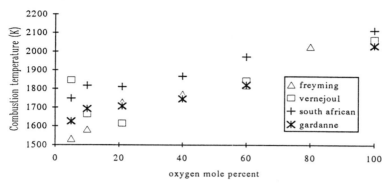

Figure 16b : Heterogeneous combustion temperatures

We used a least square minimisation to correlate our measured values to the oxygen concentration C_0 with a power law :

$$t_d = a \cdot c_0{}^b$$

b values are summed up in table 4. They are close to -1 for the four samples studied here confirming diffusional control.

Coal	Freyming	Vernejoul	Gardanne	South African
b	-0.89	-0.94	-0.99	-0.97
standard deviation	0.80	0.76	0.75	0.86

Table 4 : Least square b values

For high oxygen molar concentration of the combustion gas, the influence of the Stephan flow can not be neglected. We used the model developed by Mulcahy and Smith to predict combustion times :

$$\tau_D = -\frac{\sigma_A}{6.D_0.\rho_0}\left(\frac{T_0}{T}\right)^{0.75} d_0{}^2 \frac{\gamma}{Ln(1-\gamma.fm)}$$

τ_D : burning time (s)
σ_A : char density (g/m³)
D_0 : diffusion coefficient (m²/s)
ρ_0 : particle density (g/m³)
T_0 : gas temperature (K)
T : boundary layer temperature (K)
d_0 : particle diameter (m)
γ : coefficient related reaction stoichiometry
fm : mass fraction

If we assume that CO is the primary combustion product, γ must be taken as -1. The relative differences between computed and measured values are given in fig 17. Predictions are quite good for Freyming and Vernejoul coals. For the two other samples, we can observe an under estimation of combustion duration. This fact must be related to the high ash content of these two coals. In the laminar conditions of our experiments, the ash layer around the particle is not eliminated and must be a limitation to the diffusion of oxygen from the bulk gas to the particle surface.

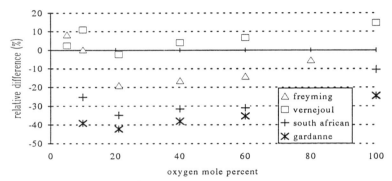

Figure 17 : Comparison between computed and measured values.

4. CONCLUSION

Coal combustion has been studied with two different reactors, the first one is a drop tube furnace which allows the study of micron sized coal particles, ignition times, flame and heterogeneous combustion duration, and combustion temperature can be measured with a three wavelength pyrometer. The influence of the oxygen content of the combustion atmosphere has been studied for five different coal samples. Results confirm the hypothesis of boundary layer diffusion control.

Two models have been developed at the laboratory in order to analyse these experimental results, they are both based on the unreacted shrinking core model. The main difference between these models is based on the observation that some coals form cenospheres during the devolatilization. The computed values agree well with the measured ones, except for 5% oxygen in combustion gas, where experimental errors may be caused by the difficulty to detect low temperatures burning chars.

The second reactor has been developed to study coal combustion with particles having diameter between 0.5 and 3 mm. The use of CO_2 lasers allow the rapid heating of the sample. A three wavelength pyrometer is used to measure combustion characteristics, and pollutant analysis can be performed by on line analysers or by batch chromatographic methods.

The influence of oxygen on combustion characteristics has been studied for four different coals. Ignition time, volatile matter flame duration, heterogeneous combustion duration and temperature have been measured. The variation of ignition delays agrees well with a transition from an homogeneous ignition to an heterogeneous mechanism when oxygen increases. A simulation based on the unreacted shrinking core model has shown that the assumption of boundary layer control was true for describing our experiments.

5. REFERENCES

1   M.A. FIELD, Comb.Flame, 1969, 13, 237
2   J.M. BEER , M.W. THRING and R.H. ESSENHIGH, Comb.Flame,1959, 3, 557
3   A.B. AYLING and I.W. SMITH, Comb.Flame,1972, 18, 173
4   R.J. TYLER and I. W. SMITH, Fuel, 1975, 54, 99
5   J. EAPEN, P.B. BLACKADAR and R.H. ESSENHIGH, 16th Symp.(Int.) on Combustion,1977, 515
6   J.L. JOHNSON, Coal Gasification. Advance in Chemistry Series n° 131,Am.Chem.Soc. , Washington,DC,1974, 145
7   S. ERGUN, Ind.Eng.Chem.,1955, 47, 2075
8   S. DUTTA and C.Y. Wen, Ind.Eng.Chem. Proc.Des.Dev.,1977, 16, 31
9   N. GARDNER, E. SAMUELS and K. WILKS, Coal Gasification. Advance in Chemistry ,1974, Series n° 131, Am.Chem.Soc. , Washington,DC, 217
10 R.H. ESSENHIGH, Chemistry of Coal Utilisation Wiley, New York, 1981, 1153
11 L.D. SMOOT, Fossil Fuel Combustion,Wiley,New York,1991, 653
12 N.M. LAURENDEAU, Prog. Energy Comb. Sci.,1978, 4, 221
13 R. BOUKARA, R. GADIOU, P. GILOT, L. DELFOSSE and G. PRADO,23th Symp.(Int.) on Combustion, Sydney,1992
14 O. CHARON,Thèse Université Haute Alsace,1988
15 R. GADIOU, Thèse Université de Haute Alsace, 1990
16 D. SCHWARTZ, Thèse Université de Haute Alsace, 1992
17 " Coal Combustion and Gasification "(L. D. Smoot and J.P. Smith Ed.) , Plenum Press ,1985
18 G.G. DE SOETE, Rev. I. F. P. ,1985, 40, 5
19 M.F.R. MULCAHY and I.W. SMITH,Rev.Pure Appl.Chem.,1969, 19, 81
20 N.M. LAURENDEAU,Prog.En.Comb.Sci.,1978, 4, 221

The Shell Coal Gasification Process: an Efficient and Clean Technology for Generating Power from Coal

J. E. Naber[1], P. J. A. Tijm[2], and M. M. G. Senden[1]
[1] KONINKLIJK/SHELL-LABORATORIUM, AMSTERDAM (SHELL RESEARCH BV), PO BOX 3003, 1003 AA AMSTERDAM, THE NETHERLANDS
[2] SHELL INTERNATIONALE PETROLEUM MAATSCHAPPIJ BV, PO BOX 162, 2501 AN THE HAGUE, THE NETHERLANDS

SUMMARY

Modern coal-gasification, combined-cycle (CGCC) power-generation technologies present important options and opportunities to electric power producers today and in the future by offering major improvements in efficiency, in environmental performance, and in overall cost effectiveness as compared to conventional coal-fired technologies. The Shell Coal Gasification Process, a coal-conversion process based on Shell's oxygen-blown entrained-flow technology, is particularly well suited to being coupled with combined-cycle systems for electric power generation. Shell's demonstration plant, SCGP-1, located near Houston, Texas, has run successfully on 18 different feedstocks covering a wide range of coal types and properties, including petroleum coke and lignite, for a total operating time of almost 15,000 hours.

The consequences of applying oxygen-blown instead of air-blown gasification for the process characteristics and economics are elucidated. The use of oxygen in CGCC technology, such as the Shell Coal Gasification Process, is essential for reasons of cost effectiveness, environmental performance, and process efficiency and scalability. Moreover, it allows the manufacture of CGCC by-products, such as methanol. Oxygen purity as optimisation parameter should of course be evaluated taking the fully integrated power plant into account.

The Shell Coal Gasification Process technology has been selected for a 250 MW coal-gasification, combined-cycle power plant in The Netherlands, targeted to start up in late 1993.

1. INTRODUCTION

Coal is the most abundantly available hydrocarbon fuel in the world, with a well-established infrastructure for its utilisation and a relatively low and stable price, which, unlike that of oil and gas, is little impacted by international events. Coal has been the primary fuel for electricity generation and is being counted on to play a continuing and increasing role as such in the future.

Conventional coal-fired electricity generation has given rise to numerous environmental problems, notably emissions of sulphur and nitrogen compounds, both of which have been linked to acid rain, and particulates emission.

Modern coal-gasification, combined-cycle (CGCC) power-generation technologies, however, present important options and opportunities to electric power producers today and in the future by offering major improvements in efficiency, in environmental performance, and in overall cost effectiveness as compared to the conventional coal-fired technologies. The Shell Coal Gasification Process (SCGP), a coal-conversion process based on Shell's oxygen-blown entrained-flow technology, is particularly well suited to being coupled/integrated with combined-cycle systems for electric power production. The use of oxygen, moreover, offers specific process advantages, as well as the opportunity to employ the produced synthesis gas (syngas) not only for power generation, but also as a chemical feedstock, as a source of hydrogen, as a fuel for heating, or for a combination of these.

2. THE SHELL COAL GASIFICATION PROCESS

The Shell Coal Gasification Process, Figure 1, is a dry-feed, entrained-flow, high-pressure, high-temperature gasification process, from which the ash is removed as a molten slag. Raw coal is pulverised in a conventional bowl mill to a size range similar to that used in conventional combustion systems. A hot inert gas dries the pulverised coal and transports it out of the mill to the coal pressurisation section and, in dense phase, to the gasifier feed system.

The oxygen required for the gasification reaction is supplied by an air separation unit (ASU). The nitrogen from this unit is compressed for use in the gasification plant, e.g. to make up inert gas for coal milling and drying, and as transport gas in the coal-feed system. Pressurised coal, oxygen and, if necessary, steam enter the gasifier through pairs of opposed burners. The gasifier (Figure 2) consists of an outer pressure vessel and an inner gasification chamber, with

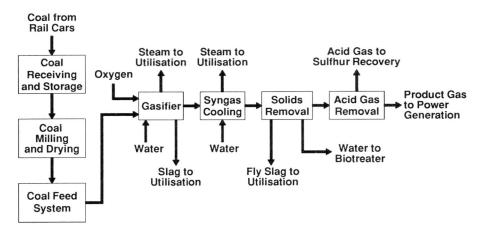

Figure 1
SCGP: Schematic Diagram

a water-cooled membrane wall. The membrane wall contains the hot reaction space, providing a high degree
of gasifier reliability. The high gasifier temperature,
typically between 1400 and 1700 °C, ensures that the
molten slag flows freely down the membrane wall into a
water-filled compartment at the bottom of the gasifier.
These high-temperature process conditions result in
high carbon conversions (above 99%) and produce a raw
syngas in which essentially no organic components
heavier than methane are present. The insulation provided by the slag layer in the gasifier minimises heat
losses, recovered by steam production, so that cold gas
efficiencies are high and CO_2 levels in the syngas are
low. Flux may be added to the coal feed to promote the
appropriate flow of molten slag.

The hot product gas leaving the gasification zone
is quenched with cooled, recycled product gas to solid-
ify entrained fly slag before it is passed to the syngas cooler. The syngas cooler recovers heat from the
quenched raw gas by generating high-pressure steam.

The fly slag contained in the gas leaving the
syngas cooler is removed using a cyclone and/or a filter. The recovered fly slag is recycled to the gasifier, enhancing the overall carbon conversion. Water-
soluble contaminants are removed from the syngas by
water scrubbing. The syngas goes to an acid-gas-removal
system, containing e.g. Sulfinol to remove the desired
amount (some >99%) of sulphur species. The acid gases
are fed to a Claus plant, where saleable elemental
sulphur is produced; to ensure maximum sulphur recovery
and minimum sulphur emissions, this plant is operated

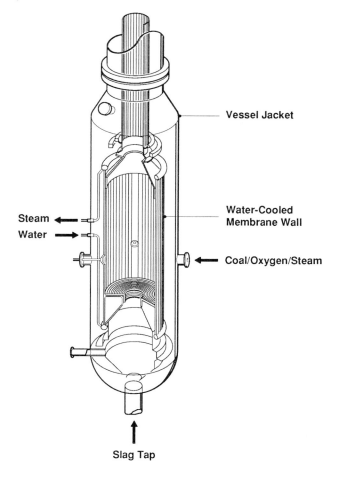

Vessel Jacket

Steam

Water

Water-Cooled
Membrane Wall

Coal/Oxygen/Steam

Slag Tap

Figure 2
The Gasifier

using the Shell Claus offgas treating process. The
clean product, a medium-BTU syngas, is routed to the
gas turbine for power generation.

3. ENVIRONMENTAL ASPECTS

The environmental performance of SCGP-based power
generation is excellent. Throughout the development of
the process priority has been given to establishing an
environmental data base and identifying design options
for treating and handling effluents.

Key environmental features of the process are the following:

- A major advantage of using coal gasification in power generation is that the gasification step converts the sulphur from the coal into reduced sulphur compounds, mainly hydrogen sulphide (H_2S), rather than into sulphur oxides, such as SO_2. The H_2S can be removed down to extremely low levels (typically some 20 ppm sulphur, representing a sulphur removal of ca. 99.5%) by gas-treating technology, which has been used in oil/natural gas industry for over 50 years.

- Nitrogen in the coal is converted chiefly to molecular nitrogen (N_2), and to some small amounts of ammonia (NH_3) and hydrogen cyanide (HCN), which are completely removed in syngas clean-up sections. Consequently, formation of NO_x from fuel-bound nitrogen is of no concern in the combustion of SCGP syngas in a gas turbine. Thermal NO_x formation is strongly influenced by the adiabatic flame temperature. Its potential can be reduced significantly by diluting the syngas with steam or nitrogen, as tests in full-scale gas turbine combustors with an SCGP-type simulated syngas have shown.

- The SCGP operates at high temperatures, which eliminates the presence of trace organics, such as tars and other heavy hydrocarbons. The mineral material contained in the coal is recovered as dense, glassy slag granules, which are excellently suited to being used in e.g. road-filling and cement-manufacturing applications.

- Particulates are removed in a dry-solids-removal section to produce a syngas meeting gas turbine particulate specifications of typically 1-5 ppm wt max.

- Environmental characterisation has shown that the process water from SCGP contains no detectable amounts of (semi)volatile organics. Biological treatment of the stripped and clarified process water provides means for the removal of remaining inorganic nitrogen and sulphur species.

4. PROCESS FEATURES AND STATE OF THE ART

The flexibility of the Shell Coal Gasification Process has been demonstrated by processing a wide variety of coals ranging from high-rank bituminous coal to low-rank lignites and petroleum coke, in an environmentally acceptable way, producing a high-purity, medium-BTU syngas, which can be used as a fuel for power generation. Seventeen different coals as well as a petroleum

TABLE 1

Summary of SCGP-1 gasification performance on the 18 feedstocks

	Carbon conversion %	Cold gas efficiency sweet gas basis % HHV
Illinois no. 5	99.7	81.6
Maple Creek	98.7	76.0
Drayton	99.6	79.3
Buckskin	99.7	78.0
Blacksville no. 2	99.7	80.5
Pyro no. 9	99.9	78.3
Sufco	99.9	81.0
Texas Lignite	99.4	80.3
Pike County (washed)	99.9	80.9
Pike County (ROM)	99.9	83.0
Dotiki	99.9	80.1
Newlands	99.7	80.3
El Cerrejon	99.6	83.4
Skyline	99.9	82.4
Robinson Creek	99.7	82.2
R&F	99.5	79.6
Pocahontas no 3	99.3	82.4
Petroleum Coke	99.5	78.9

CGE = (HHV of sweet gas)/(HHV of coal fed)

coke were successfully run at Shell's demonstration unit, SCGP-1, located in Houston, Texas; they are listed in Table 1. Over 99% of the carbon in the coal is converted. Of the energy contained in the coal being fed to the gasifier approximately 83% is converted into energy contained in the raw syngas. This ratio of energy in syngas to energy in coal is often referred to as the Cold Gas Efficiency (CGE). On a sweet syngas basis, i.e. after the removal of sulphur species, the CGE tends to be somewhat lower. With a further 16.5% of the energy in coal being recovered by steam production, the overall energy efficiency of the SCGP is very high. A typical Sankey diagram is given in Figure 3.

The composition of the syngas produced is similar for all the coals processed. CO and H_2 account for slightly over 90% of the syngas on a molar basis and the high heating value (HHV) of the sweet syngas is typically 10,800-11,200 MJ/Nm^3. With this stable composition the SCGP syngas is eminently suitable for use in gas turbines and allows for changes in quality of the coal feed to the gasifier "on the fly", as has been demonstrated in the SCGP-1 unit.

5. THE DEMKOLEC PROJECT IN THE NETHERLANDS

Sep (the Dutch Electricity Generating Board) have announced that coal gasification will play an important

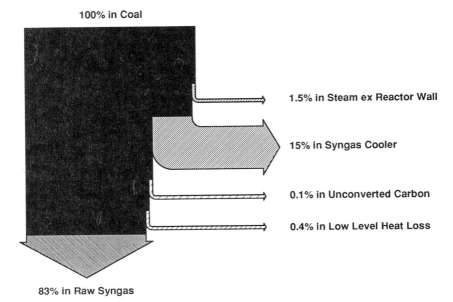

Figure 3
SCGP Energy Balance

role in their overall Dutch electricity plan. A 250 MW
integrated coal-gasification, combined-cycle (ICGCC)
demonstration plant is under construction at Buggenum,
The Netherlands. It is scheduled to start up in late
1993.

The plant is being built and will be operated by
Demkolec, a Sep subsidiary. A single gasification
train, designed for approximately 2000 t/d of a range
of imported coals, will feed a single-shaft Siemens
(V94.2) gas-steam turbine. The overall efficiency is
43.2% (low heating valve; LHV). This high efficiency is
possible through a high degree of integration, with
feed for the air-separation unit being bled from the
compressor of the gas turbine. The steam from the gasification island will be sent to the steam turbine,
together with steam from the heat-recovery steam generator. Heat exchange between the various steam cycles
has been designed to make state-of-the-art use of the
energy production, as shown in Table 2.

The Demkolec plant has been designed for very low
emissions. Table 3 shows the levels laid down in the
environmental permits, which were granted in April
1990. NO_x-reduction measures include saturation of the
syngas with steam and dilution with surplus nitrogen
from the air-separation plant.

TABLE 2

DEMKOLEC electricity production

Gas turbine	156 MW
Steam turbine	128 MW
Gross electrical production	284 MW
Internal consumption	-31 MW
Net electrical production	253 MW

Steam pressures 125, 40 and 8 bar

TABLE 3

DEMKOLEC emissions

NO_x emissions	Permit level
Gas turbine	Lower than 95 g/GJ
Total ICGCC	<75 g/GJ
Sulphur removal	At least 97.85% (for 1.5% S)
Total SO_2 emission	220 mg/kWh

6. THE ROLE OF OXYGEN IN INTEGRATED GASIFICATION, COMBINED-CYCLE PROCESSES

The use of relatively pure oxygen in SCGP is essential as special efficiency and environmental benefits are derived from it. For instance, it ensures high reaction temperatures, which enable slagging operation for any type of coal, and ensures complete conversion of carbon and volatile organic species at short residence times. Moreover, an oxygen-blown process results in smaller syngas volumes of higher calorific values than the air-blown process. Removal of trace components from the syngas is therefore simpler and cheaper for oxygen- than for air-blown processes. The high reaction temperatures result in a simple gasification diagram for the oxygen-blown entrained-flow gasifiers, comprising equilibrium gas-phase reactions and complete carbon conversion.

Combined-cycle power generation enables power producers to introduce major improvements in efficiency and cost effectiveness. Natural gas-fired integrated gasification, combined-cycle processes with overall efficiencies above 50% (LHV) are already in operation. Developments in gas turbines and steam cycles point in the direction of 55 to 60% (LHV) efficiency for such natural gas-fired plants in the late nineties.

Modern coal-gasification technology presents a unique opportunity to fuse the advantages of high-efficiency, combined-cycle power generation with those of an environmentally friendly coal-based process. SCGP, particularly well suited to producing clean synthesis gas efficiently, also provides flexibility in plant configuration. There are numerous possibilities for optimising the use of energy in SCGP-based ICGCC designs, depending on the specifics of the gas and steam turbine selected and on the site specifics.

Recent studies by ABB and Air Products (published in Modern Power Systems, November 1990) based on the use of ABB's 13E gas turbine show a state-of-the-art, optimal integration of the three main building blocks of an ICGCC plant (air separation unit, gasification block and the combined cycle). Here, the air-separation unit receives about two thirds of its air from a dedicated air compressor, with the remaining one third stemming from the air compressor of the gas turbine. The nitrogen not required in the gasification unit is used for dilution of the clean syngas fired in the gas turbine so as to reduce NO_x formation. The results, illustrated in Table 4, show that an overall efficiency of over 46% (LHV) can be achieved.

Developments of these integration schemes have provided an insight into the optimum oxygen purities for coal gasification:

- Carbon conversion capacity in coal gasification is highly dependent on the temperature of the gasification reaction and hence on the oxygen purity.

- Air-blown gasification processes have a much lower once-through carbon conversion than oxygen-blown processes and call for the recycle of material. In the end, eliminating the own energy use of the ASU, the overall process efficiency may be slightly better. However, capital investment figures indicate that the profits realised by

TABLE 4

Energy generation breakdown for a single 250 MW ICGCC train

Coal in (El Cerrejon 1770 tons/day)	539	MW (LHV)	thermal
Fuel in (natural gas)	6	MW (LHV)[1]	thermal
Total	545	MW (LHV)	thermal
Output gas turbine	171.4 MW		electrical
Available heat in steam	300	MW	thermal
Converted in steam turbine	121.5 MW		electrical
Mechanical/electrical loss	(3.7 MW)		electrical
Net steam turbine output	117.8 MW		electrical
Total production	289.2 MW		electrical
Internal consumption	(37.0 MW)		electrical
Net output	252.2 MW		electrical
Efficiency (% of LHV)	46.2		

[1] Natural gas is used as fuel in the coal milling and drying system, and in the sulphur recovery system.

eliminating the ASU will never offset the costs
resulting from the required enlargement of the
other equipment, from gasifier to Claus plant,
because of the additional nitrogen present in the
gas.

Oxygen enrichment has also been studied as a means
to improve the process efficiency and capital
investment. The impact of oxygen purity has been investigated over a wide range of oxygen concentrations. An
analysis of the overall facility cost and performance
indicated that the optimum oxygen purity was about 95%
for low-pressure, stand-alone air plants. In capital
investment terms, however, the optimum oxygen purity
has been shown to level off at 85-95% (Figure 4).
Because integrated air-separation plants extract their
air from the gas-turbine air compressor and hence are
operated at higher pressures, they are volumetrically
smaller than low-pressure stand-alone oxygen plants. As
a result, capital cost advantages can be realised for
equipment and installation and/or energy savings can be
realised by shifting the optimum purity towards 85%.

Most short-term improvements in the efficiency of
SCGP-based ICGCC stations comprise the use of heat for
the generation of chemical heat and/or the increase of

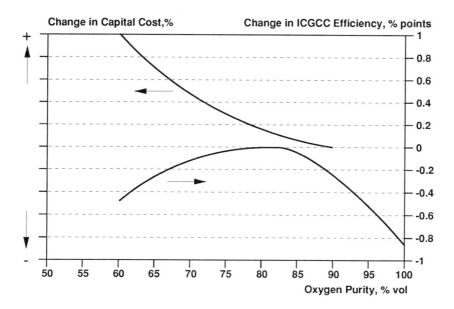

Figure 4
Effect of Oxygen Purity of the
ASU Intergrated with Gas Turbine

sensible heat of clean gas to the gas turbine. At present, heat is normally used for raising steam. A gain in efficiency results from converting gas to power with an efficiency of 50-52%, whereas steam is converted into power with an efficiency of just 30-40%, depending on the quality of the steam.

Some examples of these improvements are the following:

- Increasing the oxygen preheat temperature from 230 to 320 °C would increase the station efficiency by 0.1%. Since less oxygen would be required to produce the same amount of syngas this would result in an additional increase of 0.2%.
- Gas treating at a 100 °C higher temperature would increase the station efficiency by 0.6%. Avoiding the gas dew point is very important here.
- Using heat of the syngas cooler for preheating the clean syngas rather than generating steam would increase the station efficiency by 0.3% if the gas were heated to 500 °C. Applying the same treatment to the surplus nitrogen from the ASU would lead to an additional increase of 0.3%.

These relative improvements in efficiency can obviously be translated into reductions in capital and operational investments.

On the longer term ICGCC technology offers considerable room for additional improvements. Table 5 gives the key factors of gasification technology, together with their degree of maturity and scope for

TABLE 5

Key factors of gasification technology

Cold gas efficiency/heat rate	+++
Environmental performance	++
Capital cost	+
Availability	++
Reliability/operability	++
Novelty	++
Feedstock flexibility	+++
Materials of construction	++
Maintenance requirements	+
Construction requirements	+
Oxygen requirements	++
Combined/cycle integration	++
Scale-up/design flexibility	++
Gas purity & composition	+++
Sulphur recovery/purity	+++

+++   Sound demonstration;
++   Solid progress - not optimized;
+   Viable technology - significant gains possible.

TABLE 6
Estimated new coal-fired electric power capacity
over the period 1995-2005

Area*	Amount, GW	Annual growth rate, %
EEC	100	1.6
US	130	2.0
Rest	<u>130</u>	6.3
	360 GW	
∴ 70 x 500 MW plants per year		

*Excluding CIS and Eastern Europe.

improvement. There is a host of ideas for improving the technology, specifically from a point of view of capital cost and plant availability. Important reductions in capital expenditure are envisaged on the longer term. It will be clear that experience with the first commercial plants will significantly enhance the realisation of these improvements. This will put ICGCC technology in the right position to achieve a more than proportional share of projected new coal-fired electric power capacity (Table 6).

RELATED PUBLICATIONS (in chronological order)

M.J. van der Burgt, and J.E. Naber, "Development of the Coal Gasification Process (SCGP)", Third BOC Priestley Conference, Imperial College, London, UK, September 12-15, 1983.

R.K. Malcharek, and T. van Herwijnen, "Integrated Gasification and Combined Cycle Power Generation", Energy Production Processes Symposium, London, UK, April 1988.

T. Hope, R.K. Malcharek, and R.T. Perry, "The Shell Coal Gasification Process for Power Generation", Seventh International Conference and Exhibition on Coal Technology and Coal Economics, Amsterdam, The Netherlands, November 1988.

W.V. Bush, K.R. Loos, and P.F. Russell, "Environmental Characterization of the Shell Coal Gasification Process. I. Gaseous Effluent Streams", Fifteenth Biennial Low-Rank Fuels Symposium, St. Paul, Minnesota, USA, May 1989.

D.C. Baker, W.V. Bush, K.R. Loos, M.W. Potter, R.A. Swatloski, and P.F. Russell, "Environmental Characterization of the Shell Coal Gasification Process. II. Aqueous Effluent", Sixth Annual International Pittsburgh Coal Conference, Pittsburgh, Pennsylvania, USA, September 1989.

R.N. Franklin, R.P. Jensen, and P.J.A. Tijm, "Progress in Shell Coal Gasification", Alternate Energy '90 Conference, Carmel, California, USA, April 18-20, 1990.

J. Klosek, and J.C. Sorensen (Air Products and Chemicals, Inc., Allentown, PA), "Energy Storage with Coal Gasification Combined-Cycle Electric Power", Gulf Coast Cogeneration Association Spring Regional Conference, Houston, Texas, May 1990.

G.D. Zon, "Integration of Coal Gasification and Combined Cycle for the Production of Electricity, Demonstration Project KV-STEG Buggenum (NL)", Coal and Power Technology '90, Amsterdam, The Netherlands, 1990.

J.A. Salter, S.H. Gantz, W.T. Tang, P.J.A. Tijm, J.B. DuBois, and R.T. Perry, "Shell Coal Gasification Process: By-Products Utilization", American Coal Ash Association's Ninth International Coal Ash Utilization Symposium, January 1991.

R.P. Jensen, M.M.G. Senden, and H.L.M. Bakker, "Coal Gasification: Current Status and Prospects", World Coal Institute Conference "Coal in the Environment", London, England, April 1991.

K.D. Wiegner, P.J.A. Tijm, and F.A.M. Schrijvers, "Clean Power from the Shell Coal Gasification Process", VGB Conference 'Coal Gasification 1991', Dortmund, Germany, May 1991.

F. Eulderink, D. Chen, F.A.M. Schrijvers, and P.J.A. Tijm, "The Shell Coal Gasification Process: Today's Technology for Tomorrow's World", First International Conference on Combustion Technologies for a Clean Environment, Vilamoura, Portugal, September 1991.

U. Mahagaokar, J.N. Phillips, and A.B. Krewinghaus, "Shell's SCGP-1 Test Program-Final Overall Results", Tenth EPRI Conference on Gasification Power Plants, Palo Alto, California, USA, October 1991.

J.C. Sorensen, A.R. Smith, and M. Wong, "Cost-Effective Oxygen for GCC - Matching the Design to the Project", Tenth EPRI Conference on Gasification Power Plants, Palo Alto, California, USA, October 1991.

E.C. Heyman et al., "The Shell Coal Gasification Process-The Clean and Efficient Technology for Power from Coal", Conference Energy and Environment: Transition in Eastern Europe, Prague, Czechoslovakia, April 1992.

R. Muller, U. Schiffers, and G. Baumgartel, "Kombi-Kraftwerk mit Kohle-vergasung", VGB Kraftwerkstechnik 72 (1992), Heft 5, pp. 413-419.

Combustion Processes Part 2

The Role of Oxygen in New Steelmaking Technologies at ILVA Taranto

G. Federico, C. Liscio, G. S. Malgarini, and S. Innocenti
ILVA, CENTRO SVILUPPO MATERIALI, CORSO FM PERRONE 24/A,
16152 GENOVA, ITALY

1 INTRODUCTION

Oxygen was once used in steel production in converters to separate carbon, silica and phosphorus from iron.
Since 1972, when auxiliary fuel injection started, oxygen has been utilized in Taranto blast furnaces.
Blast enrichment is necessary to achieve a high degree of combustion of oil, tar and pulverized coal at the tuyeres.

ILVA is today studying the application of oxycoal technology at the blast furnace no. 2 in Taranto works, in order to allow higher rates of coal injection and lower coke consumptions.
A coal injection rate higher than 200 kg/t of hot metal with a blast enrichment of 90 Ncm/t is predicted to be experienced in stable operations within 1995.

A good mixing between oxygen and coal can be obtained by mean of coaxial lances for a good efficiency of combustion. Different lance geometries, with and without final swirl, and different materials will be experimented and blast furnace operational conditions will be monitored by means of special instrumentations.

Oxygen Service plant configuration at present available in ILVA Taranto allows highly reliable operations with current oxygen consumption rates and in case of oxycoal application at one of the four blast furnaces in operation; nevertheless some changes have to be predicted for further developments:

- if experimental operations at blast furnace no. 2 are succesful, the application of oxycoal will be extended to all the other blast furnaces, with an oxygen consumption increase of approximately 65,000 Ncm/h;

- beside ILVA, together with Voest Alpine, that is the licenser, is carrying out the feasibility study for the Corex process: reference for the study is the substitution of one of the consisting blast furnaces of 1.9 Mt/y of iron with a Corex production unit of the same capacity made by two modules. The associated increase is approximately 120,000 Ncm/h;

- at the same time C.C.F. (Converted Cyclon Furnace) process is under evaluation by ILVA for a later alternative to the traditional ironmaking route (coke oven plus blast furnace); it is estimated to require more or less the same amount of oxygen in Ncm/t of hot metal of Corex.

This means that in the future the oxygen production capacity
in Taranto has to be predicted in increase rapidly, with new
plant sets to ensure the same present supply reliability.

The paper shows the present air separation unit configuration
in Taranto and the predictable developments.

2   TARANTO OXYGEN SERVICE PRESENT CONFIGURATION

The present daily average consumption in Taranto work is
approximately 100,000 Ncm/h.
The hourly average consumption ranges between 95,000 and
110,000 Ncm/h, due to the variations in LD converter
consumptions, as oxygen is blown in the hot metal only during
refining but not during charging and discharging.
Gas consumption in blast furnaces is steady and it is
approximately 35,000 Ncm/h, supplied at low pressure (8 bar
abs.) in the cold blast; the remaining amount is consumed in
LD plants at 40 bar.

All the oxygen in Taranto is supplied at a purity of 99%.

Oxygen Service in Taranto includes seven operating air
separation units, split in two sections (see figure 1):

-   OSS/1, that includes two sub-sections for air separation
and air compression and distribution:

```
        Air Separation Units        Air Compressors
        (oxygen production)
- III A.S.U.   9,000 Ncm/h - 3 centrifugal x   57,000 Ncm/h
-  IV A.S.U.  16,000 Ncm/h - 1 centrifugal x   90,000 Ncm/h
-  IX A.S.U.  16,000 Ncm/h - 1 centrifugal x  100,000 Ncm/h
```

Every unit has a dedicated air compressor and can be
connected to each one of the others by manual operation.

Figure 1 Compressed air distribution

- **OSS/2** that includes 2 sub-sections for air separation and air compression and distribution:

Air Separation Unit (oxygen production)	Air compressors
- VI A.S.U. 30,000 Ncm/h	- 3 axial x 170,000 Ncm/h
- VII A.S.U. 30,000 Ncm/h	- 2 centrifugal x 100,000 Ncm/h
- VIII A.S.U. 30,000 Ncm/h	
- X A.S.U. 48.000 Ncm/h	

Automatically every air separation unit recognizes the dedicated air compressor and stops in case emergency of the compressor. Equally the air compressors discharge air automatically in case of A.S.U. trip.
The dedicated air compressor can be selected among those available.

From OSS/1 to OSS/2 compressed air can be transferred by a pipeline with manually controlled valves.

The oxygen manufactured by OSS/1 and OSS/2 is compressed in the common oxygen compression section, that includes:

- 3 centrifugal compressors at 40 bar (3x38,000 Ncm/h);
- 8 alternative compressors at 40 bar (8x4,500 Ncm/h);
- 1 centrifugal compressor at 8 bar (1x35,000 Ncm/h).

Beside two oxygen pressure reduction units (1x20,000 Ncm/h and 1x38,000 Ncm/h) are available (Figure 2).

To allow a highly reliable oxygen supply, taking into account planned maintenance periods, the plants in operation are scheduled by means of prefixed productivity configurations.
At the moment oxygen supply is ensured by two operating ASU in OSS/1 and two operating ASU in OSS/2.
In case of emergency of one operating unit, the prefixed assets (with a max. capacity of 110,000 nmc/h of oxygen) allow to start another plant even if one is already out of order for maintenance.

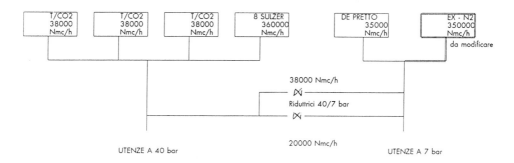

Figure 2 Oxygen compression

While the same is true for air compressors and high pressure oxygen compressors, in case of unavailability of the oxygen compressor at 8 bar, blast furnaces have to inject oxygen with pressure expanded from 40 bar: the result is a hot metal production cost penalty.

3   OXYCOAL TECHNOLOGY APPLICATION AT BLAST FURNACE no.2

Taranto blast furnaces are already equipped for pulverized coal injection (PCI): a coal grinding, drying and injecting plant has been operated since September 1991 to supply coal to the tuyeres of blast furnaces no. 1-2-4-5.

PCI plant makes available approximately 240 t/h of coal to be distributed and injected into the blast furnaces.
The nominal limit for coal injection related to the maximum hot metal productivity is 200 kg/t of hot metal. Nevertheless at the real present production rate it can be considered between 200 and 250 kg/t of hot metal.
Up to now an injection rate of 150 kg/t has been reached at blast furnaces no. 2-4, while the same value is scheduled for blast furnaces 1-5 before the end of 1992. The corresponding coke rate is around 350 kg/t: in comparison with all coke operation coal injection it allows to save a coke production of 1.2 Mt/y, this means the possibility of coke oven plant capacity decreasing that allows to avoid battery rebuilding at the end of their technical life.

In order to allow blast furnace steady operations with injection rates higher than 150 kg/t of hot metal and further coke rate decreasing, ILVA estimates oxygen enrichment in the cold blast is more effective to provide a good degree of combustion of the injected coal. Neverthless hot blast enrichment (at least for additional oxygen consumptions) is necessary.
It's necessary to have premixing of coal and oxygen without cooling down blast, as additional oxygen is injected at low temperature.

Blast furnace no. 2 will be arranged for oxygen injection at the tuyeres, while the coal inection line will be operated at the maximum capacity (50 t/h).
Mathematical analyses by means of a computer model, that takes in account operational condition of blast furnace no. 2, were made to predict corresponding oxygen and coal rates. Two reference conditions were determined:

- at a coal rate of 200 kg/t, the oxygen rate is 38 Ncm/t; coke consumption is 290 kg/t and blast is 960 Ncm/t;

- at a coal rate of 250 kg/t, the oxygen rate is 95 Ncm/t; coke consumption is 256 kg/t, and blast is 800 Ncm/h.

It has to be pointed out that oxygen increasing rate is much faster than that of coal because, beside coal combustion, an acceptable adiabatic combustion temperature (that is a thermal index for furnace operations) has to be allowed though coal is injected at room temperature and blast rate decreases.

The calculated energetic balances reveal higher electrical energy consumptions for oxygen production are balanced by less energy to heat and compress blast air and by the higher chemical energy availability in the export gas.

The advantage of oxycoal is to avoid part of the heat losses in the coke oven plant (and it means not only energy saving but also environment saving).

In order to detect the best operational condition oxygen enrichment will be made in the cold blast (before hot stoves) and also at the tuyeres, and the best distribution ratio will be researched.

Oxygen consumption will increase with coal rate and with the set adiabatic combustion temperature from the present 7,500 Ncm/h to a maximum of 22,500 Ncm/h: it means an additional maximum flow equal to 15,000 Ncm/h.

Different coaxial lance configurations will be experimented, with final swirl to allow coal and oxygen premixing and with different materials with good wear resistance at high temperature for long lance duration.

Oxygen will be supplied by a new pipeline, fed by the consisting line for cold blast enrichment at 8 bar, that will be installed within October 1993, while the experimentation period is scheduled until june 1995. The pipeline is foreseen with a global oxygen flow regulating station and with a distribution station to the single lances: the operator set oxygen flow has to be equally distributed to the operating lances.

Highly reliable safety conditions are foreseen, in order to allow the contact between oxygen and coal inside the furnace.

For a proper experimentation an instrumented tuyere, with camera and endoscope, will be installed for the monitoring of combustion and of nose condition.

Also blast furnace operations will be monitored by means of special instrumentation.

- a profilometer will be installed at the top of the furnace to check burden distribution;
- a radial gas and temperature probe will be installed above the burden for a five points gas sampling;
- vertical distribution temperature probes will utilized to provide thermal distribution from the stockline to the melting zone.

To allow a highly reliable oxygen supply with additional 15,000 Ncm/h, Taranto oxygen service has to predict the following improvements:

- a new low pressure oxygen compressor for 38,000 Ncm/h at 8 bar;
- rationalization of the process air distribution system between OSS/1 and OSS/2, to have the possibility to feed air separation units in OSS/2 with the surplus of air compressed in OSS/1. The existing pipeline from OSS/1 has to be upgraded to allow a flow of approximately 180,000 Ncm/h (with a negligible pressure drop), in order to replace one of the compressors of the other section; beside, an automatic digital system is required to manage OSS/1 compressors and to interlock compressor and air separation unit.

The new operating plant configurations will ensure a maximum hourly average consumption of 124,000 Ncm/h by mean of three operating units in OSS/2 and one operating unit in OSS/1: in this way the reliability of oxygen supply that is determinant for blast furnace thermal stability, is comparable to the present one.

4   COREX PROCESS APPLICATION IN TARANTO

The process, that is licensed by Voest Alpine Ind., takes place in two interconnected reactors (see fig. 3):

- the upper reactor (reduction shaft) that is fed by ore or pellets, fluxes descending in countercurrent with the reducing gas produced by the melter gasifier;
- the lower reactor (melter gasifier) that is supplied by oxygen, coal and ore prereduced in the reduction shaft, produces iron, slag and reducing gas.

ILVA is carrying out a feasibility study together with the licenser for Corex application in Taranto: the target of the study is a production unit including two Corex modules, for a global productivity of 1.9 Mt/y of hot metal.
In this way the Corex production unit could substitute one of the medium blast furnaces in Taranto (1-2-4) at the moment of its next relining.

Oxygen supply, whose specific consumption is more than 500 Ncm/t of hot metal, equivalent to approximately 120,000 Ncm/h, allows coal gasification to provide heat for prereduced ore melting and coal volatile matter cracking and, at the same time, to produce a highly reducing gas.

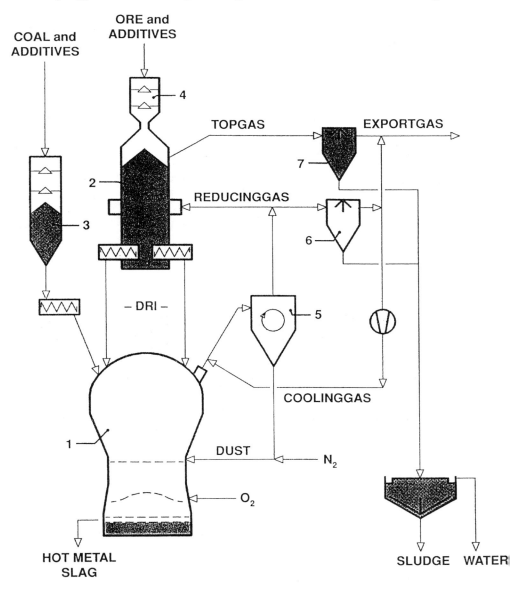

Figure 3 Corex process

The purity of 95% corresponds to technical economical optimum value, as it allows:
- acceptable coal consumption (as all the gas that is not oxygen has to be heated but does not take part in heat production);
- highly reducing characteristics for the raw gas (that depend on the percentage of carbon monoxide and water on the total);
- energy saving in oxygen manufacturing (and a slight investment cost saving for the plant).

From the point of view of the energy balance in comparison to the traditional route, Corex consumes slightly less energy, though the energy input is much higher due to coal and oxygen consumption.
The energy output in the export gas is more than the double of that of the blast furnace (also considering coke oven gas): it means that higher coal-oxygen consumptions and gas production in the Corex can balance coke-hot blast consumption and gas production in the blast furnace and that the advantage for Corex stays in avoiding energy losses in the coke oven plant.

Figure 4 resumes the previous considerations about energy balances: the two case for Corex are related to lump ore operation (case a) and pellets operation (case b, for a higher productivity).
A Corex plant has to be associated with a plant for energy recovery from export gas to be energetically and economically competitive with the traditional plant for ironmaking.

Taranto oxygen Service will have to install two additional oxygen plants, each of 65,000 Ncm/h, if Corex is installed in Taranto; it means that due to Corex the oxygen supply in Taranto will double.
Furthermore if Corex were installed and at the same time oxycoal technology were applied to the operating blast furnaces (three and not four because Corex replaces one of the existing), the average additional oxygen consumption would be 160,000 Ncm/h, to be supplied by mean of three new air separation units, each of 65,000 Ncm/h beside the modifications mentioned at the previous chapter.
The new installed plants will have to allow the maximum degree of efficiency (medium pressure plants instead of present low pressure plants) and of automation.

Also the electrical power supply should be upgraded as the power consumption increase of approximately 100 MW.
Additional electric power for oxygen manufacturing would be drawn from the existing thermal power stations, fed only by the recovery gases of the steel work without external

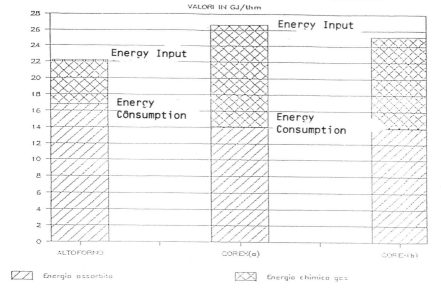

Figure 4 Comparison of energy consumptions

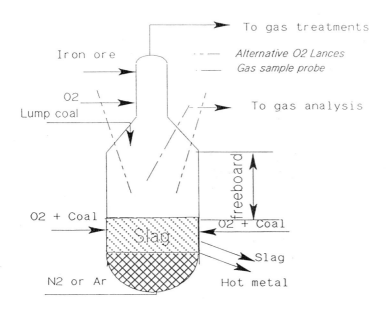

Figure 5 The CCF Process Development Unit

ancillary supply of methane or fuel oil.
New solutions integrating oxygen plant, ironmaking plant and
thermal power plant are being investigated to increse global
efficiency without affecting reliability.

6   CYCLONE CONVERTED FURNACE (C.C.F.) APPLICATION

C.C.F. is a continous process in one single reactor, with
two main components:

- the bottom part (melter vessel) where coal is gasified by
oxygen, melted ore reduction is completed and hot metal is
collected before casting;
- the upper part (cyclone) where fine ore is injected with an
angular momentum with respect the cyclone axis and it is
airborne by the ascending gas. Descending ore is prereduced
and melted.

Oxygen is injected at three different levels:

- primary oxygen: via tuyeres at a predicted flow rate of
250 Ncm/t of hot metal;
- secondary oxygen: via two lances at the top of the smelter
at a flow rate of 200 Ncm/t of hot metal;
- tertiary oxygen: via injectors in the cyclone at a flow
rate of 300 Ncm/t of hot metal.

An oxygen purity of 95% is required at a pressure of 8 bar.

ILVA and Centro Sviluppo Materiali, after having taken part
at the C.C.F. pilot plant work together with British Steel
and Hoogovens, aim to install a Process Development Unit
(PDU) designed for pressure operation(Figure 5).
The plant will be installed in Taranto and it should be
predicted for a productivity of 5 t/h of hot metal.

In order to test the PDU, trial campaigns are foreseen for
the duration of 5 days every 1 or 2 months. The uncontinous
oxygen consumption during trials will allow to utilize in
the PDU the margins of normal productivity without
modifications.
If the trials are succesful (1 year of trial since the
1994) C.C.F. could substitute one of the existing blast
furnace and oxygen plant set modification would be the same
of the previous chapter for Corex.

7 CONCLUSIONS

Present machinery availability for oxygen production in Taranto allows a highly reliable oxygen supply to LD converters and blast furnaces.

Future technologies for ironmaking, such as oxycoal, Corex and C.C.F. are being studied because they permit some slight operational cost saving and to avoid coke manufacturing that so heavily affects environment.

All these tecnologies request oxygen to allow an efficient combustion of coal and make necessary further studies for burner geometry and burner materials and to increase oxygen productivity seeking the best solutions for efficiency, automation and reliability.

Ironmaking and Steelmaking Developments

M. J. Corbett and R. B. Smith

BRITISH STEEL, TECHNICAL TEESSIDE LABORATORIES, PO BOX I I,
GRANGETOWN, MIDDLESBROUGH, CLEVELAND TS6 6UB, UK

1   INTRODUCTION

"Ironmaking and Steelmaking Developments" is a broad field and in the context of this Conference it is neither appropriate nor possible to describe in detail all of the technical developments in the iron and steel industry. Rather the intention is to present some of the perceived challenges to the iron and steel businesses of the World, to outline some key technical developments involving the use of oxygen which are ongoing at this time and to suggest areas of opportunity to promote these developments and the associated use of oxygen by the establishment of synergies between the ferrous metallurgical and tonnage oxygen industries.

Much of the paper is about ironmaking, as this is the end of the business where most of the energy and raw materials costs are incurred and where the opportunities for cost saving by process improvements are readily perceived. There are however also significant developments involving innovative uses of oxygen in steelmaking in the pipeline and some of these are also discussed.

Figures 1 and 2 show the main iron and steelmaking routes currently in use.

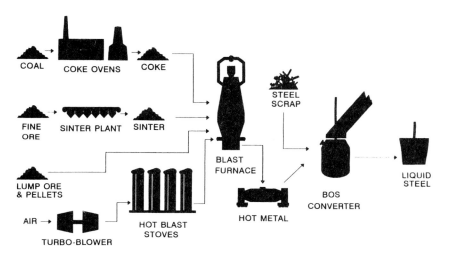

Figure 1   THE BLAST FURNACE/BOS ROUTE

By far the largest proportion of steel production (57% worldwide, 69% in the European Community) is by the blast furnace/basic oxygen steelmaking (BF/BOS) route, (Figure 1), which uses as its prime raw material and energy inputs iron ore and coal. The ores are largely agglomerated by sintering and/or pelletising and the coals are carbonised to make coke which together with the ferrous agglomerates is charged to the blast furnace. The liquid iron from the blast furnace is refined to steel in large BOS converters, in which high purity, high pressure oxygen is jetted into iron at supersonic velocities, to oxidise out carbon and metalloids. The BF/BOS route has low operating costs when worked at full capacity, it can produce very high quality products, but it is very capital intensive and is therefore usually installed at large integrated works which are not easily able to accommodate demand fluctuations.

The next most important route to steel is by melting of steel scrap in the electric arc furnace (Figure 2), which is used for the production of engineering and alloy steels, and has also been the basis of mini-mill operations producing general purpose steels often in the less demanding grades.

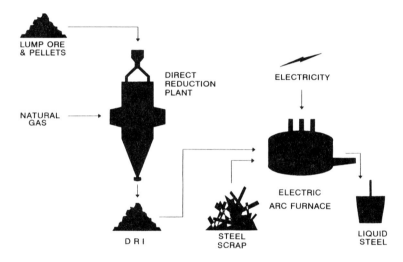

Figure 2 THE ELECTRIC ARC FURNACE ROUTE

Figure 2 also shows a variant of the electric arc furnace process in which the metallic feedstock is direct reduced iron (DRI). This route has been successfully installed in locations where the direct reduction process can use as its fuel low cost natural gas available as a by-product of petroleum extraction operations. However the cost of natural gas in most non petroleum rich economies precludes the use of this process route.

2   THE CHALLENGES

The modern blast furnace ironworks is a large, capital intensive, inflexible complex which when fully loaded can produce iron at low operating cost, but which requires regular, large injections of capital money to reline the furnaces and rebuild coke ovens and sinter plants. Figure 3 shows the proportions of the capital cost of a typical integrated ironworks distributed between the blast furnaces, coke ovens and sinter plant, and it can be seen that the investment in plant for preparation of the coke and

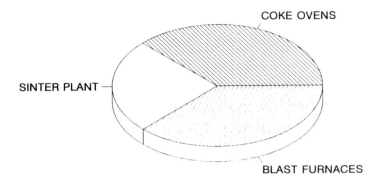

Figure 3   IRONWORKS CAPITAL COSTS

ferrous burden substantially exceeds the cost of the blast furnaces. Because of the high capital intensity the trend has been to concentrate ironmaking on fewer, larger sites to achieve economies of scale, and to operate these sites intensively to cover their high fixed costs. Clearly this strategy makes it difficult in the short term to respond flexibly to demand variations, and in the medium to long term constrains the ability to rationalise operations between several such sites.

Increasingly stringent environmental protection measures are currently being enforced worldwide, with restrictions on emissions of sulphur, nitrogen oxides, particulate matter and fume, together with fiscal penalties on carbon dioxide discharges, as key issues for iron and steel businesses. In addition to their high inherent capital cost there will, therefore, be considerable additional costs to be borne in rebuilding coke ovens and sinter plants, to enable them to achieve compliance with emission control regulations.

Most of the iron and steel businesses of the developed world are committed, by virtue of their existing investment in relatively modern furnaces, to the blast furnace ironmaking route in the short to medium term. They are many of them however facing, with varying degrees of urgency, a need for capital investment on new or rebuilt coke ovens, and the environmental pressures are expected to strengthen during the next few years, requiring expenditure to clean up sinter plant emissions. Additionally, in the face of recession and world over capacity in steel products their long term strategic business plans may, in some cases, call for rationalisation of production operations on a smaller number of core sites, which may require some incremental expansion of capacity at such sites.

In the longer term there is a need for an iron and steel manufacturing route which does not use the blast furnace as the primary smelting stage, which has a lower capital investment cost, which can therefore be viable at a smaller installed capacity, can respond flexibly to variations in the market, can fit within existing integrated Works' configurations and so is suited to staged incremental capacity changes.

In the developed economies of Western Europe and North America there is a considerable scrap arising, which offers potential energy and economic advantages as a feedstock for steel, since it can be reprocessed without the energy requirement for reduction of iron from its oxides. There are also environmental considerations which will increasingly give rise to pressures to recycle ferrous wastes Scrap prices have

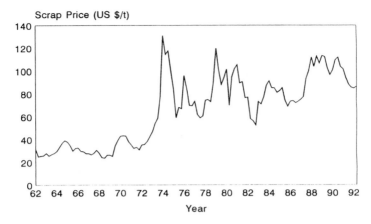

Figure 4   NO. 1 HEAVY MELTING SCRAP PRICES

historically been very volatile under demand variations (Figure 4) and there may be a two to one price fluctuation through the economic cycle. It is therefore of interest to steel producers to be able to respond flexibly to scrap price by incorporating a large proportion in the raw materials mix in times of weak demand, but at the same time not to be critically dependent on scrap during high demand, high price periods. The conventional routes to steel do not afford this flexibility, and technical developments which can improve the responsiveness to the scrap market, and indeed thereby help to regulate scrap price, are an important objective.

3   STEELMAKING DEVELOPMENTS

Fuel Assisted Electric Arc Furnace Melting

The electric arc furnace provides an alternative route to steel which is less capital intensive than the blast furnace/BOS route, does not use coke or sinter and uses cold metallic feedstocks, principally scrap. Its operating costs are however high, because of its dependence on electrical energy and the relatively high cost of carbon electrodes. The ability to use relatively cheap coal, oil or natural gas to supplement the electricity is seen as a very important development of the EAF route.

In order to attain acceptable efficiencies in steel making applications, where the product temperature is of the order of 1600-1700°C, it is necessary, when using fossil fuels, to ensure a high combustion intensity and high flame temperature. This is readily achieved by the use of pure oxygen as the oxidant in suitably designed burners.

In the U.K. British Steel, Technical and United Engineering Steels have developed together arc furnace operating practices in which oxy/coal and oxy/natural gas burners introduced through ports in the furnace sidewalls are used in the early part of the furnace cycle to supplement the electrical energy input. This has allowed up to 45 kWh/t savings on electrical energy for a fuel input of approximately 6.2 GJ/tls, and oxygen consumption of around 9.5 Nm³/t.

The Energy Optimising Furnace

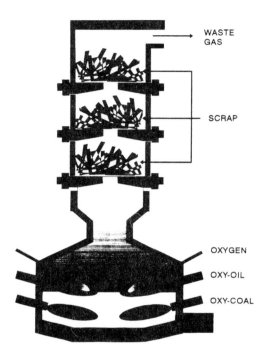

Figure 5   THE ENERGY OPTIMISING FURNACE

The energy optimising furnace (EOF) developed by Korf Lurgi Stahl is the logical development of fuel-assisted arc furnace operation. It employs a melting furnace which is similar in shape to, and has many features of the electric arc furnace (Figure 5). It does not however use electrical energy, the main heat input being from submerged coal/oxygen tuyères injecting into the metal pool. It has supplementary oil/oxygen burners firing through the sidewalls, and additional oxygen is infiltrated above the bath to combust carbon monoxide to carbon dioxide. The furnace offgases pass out through a series of scrap preheating chambers in which scrap rests on water-cooled, retractable 'fingers' in the bottom of each chamber. Scrap is charged cold into the upper chamber and then by manipulation of the fingers is caused to fall through the lower chambers one by one, before finally falling into the furnace proper. This achieves a semi counter current action in which the scrap is preheated to up to 950°C by the furnace offgases.

The EOF can use a mixed hot metal and scrap charge, or up to 100% scrap, and it therefore features good flexibility. The process appears to have considerable potential for replacement of open hearth furnaces, which in 1990 still accounted for 15% of world steel production, but could also be considered as a means of giving scrap flexibility and modest incremental capacity increases in integrated BF/BOS Works.

Enhanced Scrap Usage in the BOS

In the BOS steelmaking process it is necessary to provide some coolant in the converter to absorb the heat released by the oxidation of metalloids and carbon in refining the iron charge. This coolant is usually scrap, although other coolants such as iron ore are technically feasible. Scrap is the cheapest and most productive coolant available and contributes to the final product iron content. The amount of scrap consumed is typically in the range 200-300 kg/tonne liquid steel (tls) which can, in most steelworks, be largely or wholly met from internal revert scrap from downstream finishing operations.

If the level of scrap used in the BOS could be increased the flexibility of the mainstream steelmaking route would be significantly improved. The cost advantages available at times of cheap scrap availability could be exploited. Since the process can use other coolants or could revert to current practice with internal scrap arisings only, it would not be a hostage to high scrap prices in times of buoyant demand and could indeed probably be used as a regulator of scrap price.

In normal BOS practice the gas leaving the vessel is typically about 90% carbon monoxide and 10% carbon dioxide. The ratio of carbon dioxide to the total carbon oxides $[CO_2/(CO+CO_2)]$ is usually expressed as a percentage and called the post combustion ratio (PCR), in this case 10%. The chemical heat release in carrying out this level of oxidation of the carbon contained in the initial hot metal is about 35% of the theoretical heat available by full oxidation of the carbon. If the PCR could be raised from current typical values of around 10% to about 30%, by controlled injection of additional oxygen, it would result in a release of an additional 16% of the chemical energy in the hot metal carbon. Recovered efficiently into the steelmaking process this would provide enough sensible heat to melt an additional 60-80 kg scrap/tls.

This potential to increase scrap consumption in the BOS by raising the PCR has been well known for a long time and various attempts have been made to take advantage of it in production plants, usually by the injection of secondary oxygen into the freeboard gas space above the steel and slag in the convertor. These have not in general been successfully developed to commercial application. The reasons for this are unclear but there are known to have been problems in achieving the necessary heat recovery efficiency and, related to this, difficulties in maintaining satisfactory engineering performance of the oxygen lances and vessel refractories in the face of the increased high temperature heat flux involved.

Recent developments of coal based ironmaking processes (of which more later) have identified high levels of PCR and heat recovery as key issues in smelting reduction process and reactor design, and there are a number of reported successes in achieving them in this context. At British Steel, Technical we have been able to achieve consistent and controlled levels of around 30% PCR, coupled with heat transfer efficiencies of the order of 85%, by controlled low velocity secondary oxygen addition in pilot plant operations. If this can be demonstrated to be feasible at production scale it will provide the means of achieving the flexibility in scrap consumption which has been highlighted as a very desirable objective. Additional oxygen consumption is the key to this development.

4   IRONMAKING DEVELOPMENTS

Blast Furnace Coal Injection

Over the last ten years it has been increasingly recognised that the use of coke as the fuel in ironmaking carries significant penalties in terms of the capital cost, energy inefficiency and environmental pollution associated with coke ovens and their operation. In response to the perceived problems, and in anticipation of very heavy expenditure when existing coke ovens become due for replacement, iron and steel manufacturing businesses have sought alternative fuels for use in blast furnaces. European companies have led the way in this and the principal alternative fuel which has been chosen to replace coke has been uncarbonised coal.

In order to increase the levels of coal injection which can be applied in the blast furnace it is necessary to provide a means of maintaining raceway flame temperature, which would otherwise be depressed by the injection of cold fuel in the combustion zones and by the energy absorption due to dissociation of the coal volatiles. This temperature must be high enough to ensure complete and efficient combustion of the coal, to avoid carry-over of carbon in the furnace topgas and to prevent operational instability.
The use of oxygen to replace part of the air blast has long been known as a means of increasing the raceway flame temperature in blast furnaces. However it is only fairly recently (within the last 5-10 years) that process and engineering improvements in the technology of tonnage oxygen production by cryogenic air separation have allowed the price and availability of oxygen to make it an economically viable and practically feasible option to use it in large volumes in ironmaking.

Replacement of part of the air blast with oxygen is now, however, becoming recognised as a crucial development to allow further increases in blast furnace coal injection rates, from the currently proven and accepted operational levels of 180-200 kg/thm to anticipated rates of up to 300 kg/thm which will be needed by many companies to avoid otherwise inevitable coke oven rebuilds.

Coal injection in the blast furnace has been developed within British Steel principally at Scunthorpe Works, where injection rates of 160-180 kg/thm are now routine. This level of injection is supported by some oxygen enrichment of the air blast to maintain the necessary raceway flame temperature at the tuyères. The level of enrichment depends on the available hot blast temperature, and at Scunthorpe would typically be 6-8% with blast temperatures of the order of 1100-1140°C.

To reach substantially higher levels of coal injection the technique known as oxy/coal has been developed in which individually metered oxygen supplies are provided via injection lances at each tuyère. This avoids the need to pass highly enriched blast through the stoves and hot blast main, which could be hazardous, and provides a means of local flame temperature control in the event of variations in coal injection rates at different tuyères.

A collaborative project between British Steel, Hoogovens and Ilva/CSM with European Coal and Steel Community support was started in 1988 with the objective of demonstrating higher levels of coal injection than had hitherto been achieved using the oxy/coal method. This involved pilot plant testwork in which coal rates of over 400 kg/thm and coke rates below 200 kg/thm were demonstrated at Teesside Laboratories, followed by three production plant trials on the No.4 blast furnace at Cleveland Iron, in which coal rates up to 300 kg/thm and coke rates down to 270 kg/thm were achieved. A final trial under this Project is currently planned to demonstrate furnace operation with coke rates down to 250 kg/thm or below on a sustainable basis.

In trials carried out at British Steel's Scunthorpe Works coal injection rates in excess of 200 kg/thm have been demonstrated to be operationally sustainable in the longer term, using both blast enrichment and oxy/coal techniques for flame temperature control. This level of coal injection is in line with best practices world wide and is in fact believed to have set a world record at the time.

The current state of development of coal injection would allow coal rates up to 200 kg/thm to be specified with confidence for a routine operation, using either blast enrichment or oxy/coal for the extra oxygen required. To be able to specify 250 kg/thm would need some further development work and an extended trial period to confirm the feasibility. Operation at this level would probably need to use the oxy/coal technique Higher levels of coal injection, up to or beyond 300 kg/thm, may well be feasible in the longer term but planning for an extended demonstration of such an operation would be contingent on success at the intermediate, 250 kg/thm stage.

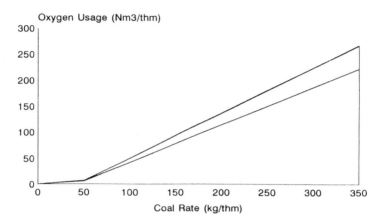

Oxygen Usage (Nm3/thm)

Coal Rate (kg/thm)

Figure 6 OXYGEN USAGE v. COAL RATE

An essential feature of all the developments aiming towards the increasing use of uncarbonised coal in blast furnaces is the use of progressively larger quantities of oxygen to support the combustion of the coal. Figure 6 shows the oxygen requirement per tonne of hot metal as a function of the coal rate, covering the levels of coal injection which are already a feasible industrial reality and projecting forward to the very high levels which form the long term objective and are currently under experimental development. The spread on the projected oxygen requirement reflects variations in the operational characteristics such as blast temperature and aim flame temperature between different plants, but also in the more futuristic projections the technical uncertainties which remain to be resolved by experimentation and plant trials.

The main scope for cost saving using coal injection arises from its potential to provide a means of avoiding a proportion of the capital expenditure which would otherwise be needed when coke ovens require replacement. For example in this circumstance installation of a coal preparation and injection plant to permit an injection rate of 200 kg/thm in a conventional ironworks would reduce the coke requirement by about 180 kg/thm.

However for high levels of coal injection to fulfil their maximum potential for cost reduction in ironmaking it will be increasingly important that oxygen is available to the ironworks in the necessary quantities and at the right price. This subject will be discussed again in a later section, but it should be emphasised that oxygen cost and availability are the keys to promote the application of coal injection technology.

Blast Furnace Fine Ore Injection

There is starting to emerge a development strategy to use fine iron ore concentrates directly in the blast furnace, by injection through the tuyères. As with coal injection the use of oxygen to maintain flame temperature in the raceway by counteracting the cooling effect of the injectant is a key feature. This development has been in response to a number of different perceived opportunities.

The most immediately apparent of these is reduction of ferrous raw materials cost which can be illustrated by reference to Figure 7.

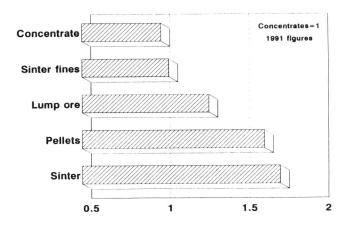

Figure 7   FERROUS RAW MATERIALS COSTS

The conventional blast furnace requires its top-charged ferrous feedstocks to be presented in a lumpy form, essentially free of fine material, to allow good access of gas to the iron ore to be smelted in the furnace. Modern practices are largely based on high purity concentrates agglomerated by sintering or pelletising. These processes, which are carried out at high temperatures, impose energy, environmental, capital and operating cost penalties on the conventional through process route.

Iron ore pellets are, for the most part, manufactured by mine operators close to the mine site and sold for shipment to the end user steel companies. Lumpy ores are handled in much the same way. There is a short term opportunity to reduce costs by displacing purchased pellets and lumpy ores by directly injected fine ore concentrates.

Sinter, which forms the majority of the ferrous burden in most modern blast furnaces, is unstable and degrades on storage and in transportation. It is therefore normally produced by the iron producer close to the point of use. The process, which is carried out at high temperature, consumes energy, and the sinter plant represents a significant capital investment. The opportunity to exploit the potential of ore fines injection to achieve cost savings by replacement of sinter arises to its full extent, either when an existing sinter plant reaches the end of its useful life, or requires additional expenditure to meet environmental protection regulations, at which time the implementation of the alternative technology allows avoidance of further capital investment in modification or replacement capacity.

There may also be other, Works-specific situations where fines injection may give immediate cost benefits, for example where the iron output is constrained by sinter availability or ferrous charging capacity. There may also be opportunities where a reduction in the sinter plant output could allow either a cheaper blend of ferrous materials to be used in the sinter feed mix, sinter quality to be improved, or the working life of the plant to be extended. These would have to be evaluated for the individual Works' circumstances.

Other potential advantages and opportunities arising from blast furnace ore fines injection are summarised in Figure 8.

- RAW MATERIALS COST SAVINGS

- IMPROVED COAL INJECTION CAPABILITY

- CAPITAL COST AVOIDANCE

- GREATER PURCHASING FLEXIBILITY

- ENHANCED HOT METAL PRODUCT QUALITY

- COST SAVINGS IN STEELMAKING

- LESS ENVIRONMENTAL POLLUTION

- ENHANCED PRODUCTIVITY

Figure 8 BLAST FURNACE FINES INJECTION OPPORTUNITIES

A further spur to the application of fine ore injection is provided by its synergy with high levels of coal injection. It is recognised that as coke is replaced by coal in the blast furnace the ratio of ferrous material to coke in the furnace stack increases and this may lead to a reduction of the furnace permeability to gas passage which could result in poor driving and instability. As the injection of fine ore will reduce the proportion of the ferrous input charged into the furnace top it will, when used in conjunction with coal injection, help to redress the imbalance in the stack ore to coke ratio and so promote improved furnace operation. As a direct consequence the process will be less vulnerable to adventitious operational disturbances, which it has already been observed can result in loss of stability, because the furnace will have a greater inherent thermal reserve. This is seen as an important feature in permitting the level of coke replacement with coal to be further extended.

In pilot plant trials it was observed that the hot metal silicon content was substantially reduced as the tuyère ore injection rate was increased.

If this can be confirmed at the production scale it will facilitate reduction in steelmaking costs and, by minimising the need for intermediate treatments, should further contribute to improved through process energy efficiency. The potential through-process cost advantages, although difficult to quantify with any precision in advance of planned plant trials, could be one of the most significant benefits accruing from fine ore injection.

It is expected that the use of fine ore/coal injection, coupled with oxygen to maintain flame temperature and promote assimilation of the injectants will facilitate increased blast furnace productivity. The replacement of hot blast with oxygen reduces the overall specific gas volumes per unit of production in the bosh and stack of the furnace, due to the decreased nitrogen content of these gases. In other words for a given bosh and stack gas flowrate and velocity the fuel burning rate, heat release and reduction/melting capacity are all greater with more oxygen and less air in the blast.

The process limit on blast furnace driving rate is imposed by considerations of gas velocity and pressure drop, which if they exceed critical values lead to mal-operation due to falling gas efficiency, excessive blowing pressure, incipient stockline fluidisation and inadequate liquids drainage (the symptoms of overblowing). Because of the decrease in nitrogen in the blast when ore/fines/coal injection is applied, these limiting parameters can be maintained below their critical values whilst the furnace output is increased. The limit on ferrous charging capacity will no longer be a determinant of output, since the extra ferrous input will be injected through the tuyères, as will the additional fuel to smelt the extra iron. The adverse effects of furnace pressure drop will additionally be ameliorated by the inclusion of a larger proportion of coke in the stock column giving improved permeability. It will still be necessary to ensure that the frontside can sustain the increased volume of hot metal, but the costs involved in modifications to permit this will be modest in comparison to those entailed by rebuilding the furnace to a bigger working volume.

In a time of depressed market activity and structural overcapacity in the iron and steel industry a general increase in furnace productivity may seem inappropriate but, ore/coal/oxygen injection, selectively applied at geographically favoured and modern furnaces/sites to increase output without major capital investment, should provide the opportunity to rationalise production on a smaller number of furnaces to meet existing demand.

British Steel Technical has carried out trials on its pilot blast furnace at Teesside Laboratories, in which fine ore injection rates up to 750 kg/thm, corresponding approximately to half of the total ferrous input, have been successfully demonstrated. Co-injection of coal and fine ore has also been tested on the pilot plant, at ore rates over 400 kg/thm and coal rates over 200 kg/thm, under which conditions the furnace permeability and stability were in fact improved over the base operation.

More recently it has been reported that Sumitomo Metal Industries has been testing co-injection of fine ore and coal on a single tuyère of a furnace at its Wakayama Works, and may have achieved local injection rates of up to 200 kg/thm of each injectant simultaneously. It is not clear whether Sumitomo has any plans to extend this to a larger trial or production operations at this time.

As with coal injection the use of oxygen is a central feature of the use of directly injected iron ore fines. Figure 9 shows the anticipated ranges of oxygen usage in the blast furnace to support ore fines injection up to a level of 600 kg/thm, with and without coal injection at 200 kg/thm.

Again there is a clear case that for fine ore injection to fulfil its maximum potential for cost reduction in ironmaking it will be increasingly important that oxygen is available to the ironworks in the necessary quantities and at the right price. As with coal injection, oxygen cost and availability are the keys to promote the application of the technology.

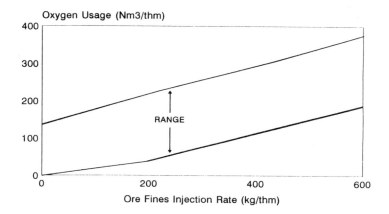

Figure 9 OXYGEN USAGE v. ORE INJECTION RATE

The New Coal Based Ironmaking Processes

There is considerable worldwide interest in, and development activity aimed at, processes to make liquid iron without using the blast furnace. The objectives and opportunities to which these new processes are aimed are shown in Figure 10. These

- ● **LOWER CAPITAL COST**

- ● **SMALLER VIABLE PLANT CAPACITY**

- ● **RAW MATERIALS COST SAVINGS**

- ● **OPERATIONAL & COMMERCIAL FLEXIBILITY**

- ● **LESS ENVIRONMENTAL POLLUTION**

- ● **ABILITY TO INTEGRATE WITH EXISTING WORKS**

Figure 10 COAL BASED IRONMAKING OBJECTIVES

processes are in general of lower capital intensity than the conventional blast furnace ironworks complex, can be viable at lower production capacity, are potentially more energy efficient and are environmentally superior. They are therefore well suited to establishment of a limited new manufacturing capability under greenfield site conditions. Because they make a liquid iron product similar in many respects to blast furnace hot metal, they could also have a rôle in providing incremental expansion or replacement capacity as may be needed to allow rationalisation of production operations across existing integrated Works based on the BF/BOS route. As such they provide a technological opportunity to evolve from a manufacturing base exclusively founded on blast furnace ironmaking to one in which the dependence on coke ovens, sinter plants and blast furnaces can be progressively reduced with a minimum redundancy of existing investment.

Attempts to develop ironmaking processes which could use 100% uncarbonised coal as the fuel and 100% unagglomerated ore as the ferrous feedstock are not new. Until the last ten years such efforts had not achieved any measure of commercial success, due to a combination of factors such as the economic pre-eminence of the blast furnace, non-availability of the necessary engineering technologies, and lack of cost and environmental incentives to pursue their application. However, from about 1980 onwards, the problems for the blast furnace came increasingly to be recognised, and all around the world new process developments aimed at realising these objectives have started up. A clear trend has emerged more recently for the new ironmaking process developments to be carried out by multi-company consortia, as the best strategy to spread the considerable development costs and risks across a sufficiently broad potential operator base.

The new ironmaking processes are in general configured to carry out prereduction/preheating and final reduction/melting in separate reactors. The initial prereduction and heating is typically effected using the hot reducing gases generated in the final smelting stage, to which the prereduced material is fed. In some variants the reduction reactor takes the form of a counter current, fixed-bed, shaft reactor, in which case the ferrous material is required to be in a lumpy form as for the blast furnace, but there is also a group of processes in which prereduction of fine ore can take place without agglomeration.

There is at the present time a relatively small number of very significant developments ongoing which should be mentioned briefly (Figure 11).

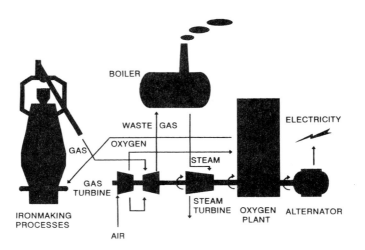

Figure 11 THE NEW IRONMAKING PROCESSES

COREX is a process based on smelting reduction, with coal as the fuel and pure oxygen as the oxidant. Unlike later generations of new processes it can only operate with lump or agglomerated ores. It was originally developed by Korf Engineering, which built and operated a large pilot plant at Kehl-am-Rhein, and it is now marketed by Deutsche Voest-Alpine of Germany and Voest-Alpine of Austria. A production scale unit of 300 ktpa capacity has operated at the Pretoria Works of ISCOR in South Africa since late 1989, and there are other projects at a pre-contract, feasibility study stage. COREX is the only new ironmaking process approaching commercialisation at this stage.

AISI - The American Iron and Steel Institute is co-ordinating a large pilot/demonstration project with participation of the main North American steel businesses and also Universities and Research Institutions. The main pilot plant is situated at the U.S. Steel site at Pittsburgh. The process is now configured to use raw coal and oxygen with agglomerated ore feedstocks. It is intended to provide a technological springboard to stimulate the revitalisation of the U.S. iron and steel industry, which has suffered from many years of under-investment.

DIOS (Direct Iron Ore Smelting) is a Japanese National Project, promoted by MITI as financier and catalyst, in which eight of the big Japanese iron and steel businesses are collaborating. It aims to use unagglomerated fine ore, coal and pure oxygen. A seven year programme, which started in April 1988, is budgeted at B¥13 of which B¥10 is allocated by MITI and the balance is provided by the steel companies via the Japan Iron and Steel Confederation. The project is currently at the stage of a large pilot plant (500 t/d) which is under construction at the Keihin Works of NKK.

HISmelt (High Intensity Smelting) is a smelting reduction process under development by CRA and Kobelco/Midrex, who have announced their intention to build a demonstration plant at Kwinana in Western Australia. In the early days of its development it was pioneered by CRA and Klöckner, but the latter company subsequently withdrew from the project. The HISmelt process is intended to operate at elevated pressure, but unlike the other forerunner developments it will use hot air blast rather than oxygen as the oxidant.

Jupiter is a new process concept under development by Thyssen, Usinor/Sacilor and IRSID as a joint venture. The concept involves the use of uncarbonised coal, fine iron ore and oxygen with a recirculating fluidised bed pre-reduction unit, and also incorporates gas recycle using physical and chemical cleaning to allow the process offgases to be recycled, in the interests of improved energy utilisation. This Project has been granted European Coal and Steel Community (ECSC) financial support under the Iron and Steel Pilot and Demonstration Programme.

CBF (Converted Blast Furnace) was a process concept originated and developed by British Steel and Hoogovens. The basis for this process was the modification of an existing blast furnace by replacement of the bosh and hearth by a continuous, coal based melting reactor, and the retention of the upper part of the stack as a direct reduction shaft. Reducing gas generated in the melting reactor would be passed to the reduction shaft to prereduce and preheat agglomerated ferrous material which was then fed to the melter unit. Extensive process and engineering feasibility studies were carried out and confirmed the technical viability of the concept. An effort was made to form a transnational European consortium to promote the development, but at the time an alternative concept, CCF (see below), was considered superior and the effort and finance were transferred to the development of this.

CCF (Cyclone Converter Furnace) is a process which has been under development by a consortium of British Steel Technical, Hoogovens and Ilva/CSM since 1989. It grew out of the earlier CBF concept but is now considered as a new process in its own right, with no requirement for a blast furnace. It aims to use fine ore, uncarbonised coal and oxygen to produce a liquid iron product in a two stage process with a final iron bath melting reactor and a preheating and reduction stage. The melting reactor is an iron bath unit in which coal and oxygen are introduced via top lances. The preheater is based on a cyclone reactor, fired with the melter offgas and further oxygen and fed with fine ore, which transfers a molten intermediate to the final stage.

The first phase of the development was carried out by the three partners on a shared-cost, shared-activity basis, supported by grant aid from the ECSC. It has so far involved pilot scale testing of the main process units at British Steel Technical, Teesside Laboratories, hot cyclone testwork at Hoogovens/International Flame Research Foundation at IJmuiden, together with extensive physical and mathematical modelling at CSM in Rome. The initial phase of the project ended in December 1991 and an interim phase, intended to lead to a specification for a demonstration plant, has been granted ECSC funding under the Iron and Steel Demonstration Programme.

Of the new ironmaking processes mentioned here only Corex has reached the stage of commercial deployment, and since there is only one plant in operation at the present time its ultimate potential remains unproven. All of the other processes are still a number of years from the construction of production scale plants. It is clear however that the iron and steel businesses of the developed world believe sufficiently in the long term need for, and future of, such processes as an alternative to the blast furnace, to invest in these developments and it is considered likely that the first decade of the next century will see one or more of these developments come to fruition.

As with the blast furnace injection developments, the new ironmaking processes in general will use very much more oxygen than has historically been the case perhaps up to 700 Nm^3/thm. It is clear that this is another example of a technological forward thrust which will require the economics and logistics of oxygen availability at the right price to provide the impetus for commercial realisation. It is suggested however that the oxygen cost and supply dimension is of key importance in the promotion of the new ironmaking technologies, and may prove to be a central criterion in their adoption by iron and steel businesses.

5   THE OXYGEN OPPORTUNITY

The three areas of ironmaking development described may be considered as short, medium and long-term visions of the future of ironmaking. They have in common a need for a cheap and plentiful oxygen supply to facilitate their commercial deployment and their projected oxygen requirements increase over the lengthening time horizon. Their other common feature is that they will lead to progressively greater energy arisings from the process off gases, and these gases will also be of progressively increasing calorific value.

To explain this statement the conventional blast furnace consumes a significant proportion of its own topgas in the form of energy for the turbo-blowers which provide the blast air, and as fuel for the hot blast stoves which preheat that air. As pure oxygen is used in the blast furnace the blast air requirement decreases and so the energy consumption for blast preparation also falls, leaving more arising gas available for export. At the same time because the ratio of air:oxygen in the blast is decreased so the nitrogen content of the arising gases falls and so the calorific value will increase. Finally as less coke and more coal are used in the developing processes the volatile matter and hydrogen contents of the fuel increase. These components which in conventional practices are removed as coke ovens gas in carbonising now appear in the ironmaking process offgas contributing to the arising energy. In the ultimate case of the coal and oxygen based smelting reduction processes these in general use all coal and no coke, they use no hot blast, and therefore consume no topgas for blowing and blast heating, and the topgas contains little or no nitrogen.

These developments must then address a dual new challenge of how economically to dispose of the surplus arising gas, and how to secure a cheap and plentiful oxygen supply. Each of these two challenges is also an opportunity which provides the solution to the other. In a conventional ironworks, any arising gas in excess of the Works' heating requirements is usually used to raise steam which then passes through turbo-alternators to make electrical power. The thermal to electrical efficiencies of such power generation schemes is usually quite poor (20-30%). Some of the generated power is used within the Works and the rest is sold to national power utilities, very often at poor prices. The main operating cost of oxygen production is energy for compression and refrigeration, and this is generally purchased as electricity from the national power utility.

As the ironmaking process developments are progressively applied, unless some alternative technology is employed, the ironmaker will convert progressively more gas into electricity via the energy inefficient steam cycle, and sell that electricity to utilities at very low prices. At the same time he will purchase more oxygen, made using electricity purchased from the same power utility at premium cost.

Fortunately a solution presents itself by application of well developed technology and engineering to integrate the ironmaking energy arisings and the oxygen production requirement, as shown in Figure 12. As the offgas calorific value will increase as a result of the developments it will be increasingly suitable for use as a fuel in gas turbines, which if operated in a combined cycle with boilers and steam turbines, using the gas turbine waste heat, can offer very much better thermal to mechanical efficiency (40-50%).

- ## COREX

- ## AISI (American Iron and Steel Institute)

- ## DIOS (Direct Iron Ore Smelting)

- ## HISmelt (High Intensity Smelting)

- ## Jupiter

- ## CCF (Cyclone Converter Furnace)

Figure 12   IRONMAKING ENERGY AND OXYGEN PRODUCTION
PROCESS INTEGRATION

The cryogenic air separation process needs its energy in mechanical form for compressor and refrigeration plant drives, and so the process integration is readily achieved.

Other positive features of the proposed integration are that the gas arisings and oxygen requirements will co-vary so the match between the parts of the system will therefore be assured, and in general both the ironmaking and air separation processes aim to operate at a steady rate 24 hours a day, 7 days a week.

In conclusion one can propose that the future for ironmaking lies in progressively increasing oxygen usage, that this future requires cheap oxygen in very large quantities, and that this future can be underpinned by exploitation of the synergy between arising energy and oxygen production by close integration of these process elements.

The Future for Oxygen in Steel Production

Jean Michard

USINOR SACILOR, 4 PLACE DE LA PYRAMIDE, 92070, PARIS, FRANCE

1   INTRODUCTION

Steel processing is, I think, the foremost client of
the oxygen industry. These two industries, steel-
processing and oxygen, are closely linked, and clearly
the development of one is related to the development
of the other. Steel-processing is evolving; indeed it
has always done so. I believe that this development
has been pronounced over the last forty years, a point
the general public is slow to recognise, as our
profession is sometimes seen as being behind the
times. We must take cognisance of the fact that in
the years to come new developments will be taking
place. What are the main features of this
development?

There are bound to be technological adjustments
to this and that piece of machinery, but I think it is
even more vital to realise that this development will
take place against a background which is fairly well
determined. It is pretty clear that, at least so far
as the great industrialised nations are concerned, no
increase in production is to be expected.

Even the assumption that production can be held
at a constant level is optimistic if we consider that
steel tends to be its own worst enemy, as the constant
process of refinement and up-grading of its qualities
reduces the weight of the steel to suit the function
for which it is required. That said, throughout its
production in Europe, steel-processing has to be
thought of in terms of large technical and economic
conglomerates, and the development I speak of will
comprise two major trends, which I believe to be quite
irreversible.

In the first place there is the increasing shift
towards the recycling of scrap iron. Steel, after
periods of service which vary considerably in length
depending on the uses to which it has been put,

reverts to a state of scrap-iron. For many reasons
such as environmental considerations, conservation of
energy, and indeed for economic ones, scrap-iron will
become more and more significant as a proportion of
the whole. So much so that, whereas today steel
recycled from scrap is about 35% of total steel
production as against 65% for metal derived from ore,
it is possible to say that within just twenty years
the proportions will be decisively reversed, with
steel from scrap rising to take up more than 60% of
the total. This, then, is already a fairly striking
development.

But there is a second one, namely profound
modifications that are taking place in some areas
where steel is produced, in steel's properties. Take
the trend arising from research into the uses of thin
sheet metal, such as for car production, the trend in
fact towards an ever purer form of steel. Whereas at
one time steel quality (in residual production) was
expressed in thousandths of a percentage point, we are
now increasingly into the era of ppm. The Japanese
take the view that we are already into the age of
steel purity levels of 6:9, to borrow from aluminium
parlance, and they are getting ready for a tomorrow of
9:9. This is a fundamental development. It is vital
that those responsible for the development of steel
should prepare for the consequences of this
development as they unfold in both space and time. It
needs hardly be said that this development will have
implications for the use of oxygen in the achievement
of such results.

Given that human activity will depend on steel
production for a long time to come, with parallel
advances needed both in production techniques and in
the nature of the steel products themselves, let us
look at the stages of steel production. For my part I
will try to indicate the major ways of possible
evolution, in particular to answer - what part will
oxygen play in all this?

2   THE THREE STAGES IN THE PRODUCTION OF STEEL

There are three stages in the steel process, as
summarised in Table 1:

1   The 'primary' stage which produces the crude
     liquid steel through use of the blast furnace,
     the oxygen process (Bessemer/BOP), and also on
     occasions the electric arc furnace.

2   The 'secondary' stage passes the liquid metal
     through reactors to remove noxious elements. The
     aim is to improve steel quality. Vacuum
     conditions are crucial.

3   The 'tertiary' stage brings the metal to a solid
     state before treating it thermomechanically.
     Cold-lamination gives little scope for oxygen, so
     I shall deal only with heat-lamination (hot
     rolling).

Table 1   Stages in the Production of Steel

Phase	Product	Process Units
PRIMARY METALLURGY	Crude liquid steel	Blast furnace
		Corex
		Oxygen refining process
		Electric arc furnace
SECONDARY METALLURGY	Refined liquid steel (for casting)	Pocket furnaces/converters ('minis' and 'midis')
		Vacuum treatment
		Shaping
TERTIARY METALLURGY	Hot rolled product	Forging
		Continuous casting
		Hot rolling

     What part does oxygen play in all this?
Combustion proper uses air, pre-heated to a greater or
lesser extent (it doesn't need pure oxygen) and
results in oxides of carbon and H_2O (with more or less
pronounced excess of oxygen).

     Semi-reductive combustion is rather more complex.
Professor Prado has tackled the question as to whether
combustion is the right word to use where coke is
gasified in the tuyère region of blast furnaces. This
tuyère gasification process is world-wide with oxygen
playing a 'reforming' role. CO_2 and H_2O are only in
the nozzles of the tuyères for a few seconds before
becoming CO, H_2 and N_2. The Corex makes use of this
reforming role, as do techniques which rely on

gasified reduction in a fluidised medium. Finally
oxygen helps in refining liquid steel by removing
carbon. Coal is increasingly used on a grand scale in
oxygen conversion, on a microscopic one in the
'secondary' stage where the carbon is in a volatile
state, but also in the oxidised liquid dross where
noxious elements such as phosphorus are removed. So
the uses of oxygen in steel production are various and
they get confused, but they are often remote from the
notion of combustion as such.

    Looking again at the three stages, but in reverse
order, we see the tertiary stage moving towards
tighter and tighter production chains. The wide
adoption of continuous processes, the linking of flow
and lamination processes, and the future adoption of
new continuous processes are all aimed at producing
thinner products, and all push production into a more
tightly organised chain. Cooling, storage, preheating solid metal, and the use of combustible
material will no longer happen in the future.

    Stage two is different. It comprises various
stages within it: the removal of oxidised phosphorus
and sulphur in the dross (slag); using a vacuum to
remove elements such as dispersed oxygen, carbon,
hydrogen and (some) nitrogen; and the control of
temperature whether electrically or by oxidising
aluminium. All this requires the purest oxygen and
almost a complete absence of nitrogen, as it degrades
the quality of the steel. The secondary stage gives a
very modest role to oxygen when compared with the
scale of modern production. It is at the primary
stage, with its huge shifts in heat and mass, that
oxygen is going to be used on a major scale.

The Primary Stage

    This stage consists of a group of operations
which yield steel from either scrap or ore. The
theoretical energy requirement to produce liquid iron
at 1600° from scrap is 320 thermies but from pure
Fe_2O_3 it is 2000 thermies. So, given the great
differences in thermic energy needed to produce steel
from ore as opposed to various forms of scrap, it is
clear that the planning of the primary stage is
crucially dependent on: the nature of the primary
material; the assessment of energy needs; the
complexity of the production process; and the costs of
production (and even more of investment). We need
also to distinguish direct from indirect fusion.

    Direct fusion involves a straightforward move to
the liquid steel state in an oxidising environment -
as in the traditional operation of the furnace with
the conventional arc. Direct fusion is the
appropriate means for dealing with recycled scrap.

Indirect fusion means going through a pronouncedly reductive phase to yield a carburated primary metal whose carbon will have to be removed in a later oxidising operation. This indirect fusion is the main means of processing steel from ore. There are normally two phases in this process from iron oxide state to that of carburated liquid metal: pre-reduction to a solid state where some part of the oxygen in the iron oxide is removed; and fusion reduction, which finishes off the oxygen by passing into a carburated liquid state. Fusion reduction (fusion gasification) is currently performed in the blast furnace, which itself has to use modified primary materials: agglomerated (sintered) ore, and coal converted into coke. New techniques for fusion reduction are seeing the light of day: the Corex, which does without coke, and others which rely on fluidisation and aim to bypass both coke and sintering. We will come back to this later.

There are two exceptions to the strong correlation we have seen between types of primary materials and choice of fusion method. One is that the carburated liquid phase can be bypassed where the ores are quite pure and where sulphur is absent. The solid pre-reduced product can then be melted simply in the electric furnace, for which it is in fact an ideal feed. The other exception is that indirect fusion can in some cases be appropriate for scrap, which is then melted along with carbon. The liquid casting so derived is later treated by the classic methods of steel-processing using oxygen. This vision of the future is summarised in Figure 1.

Leaving the exceptions aside, we arrive at a simple formulation:

- primary material scrap - direct fusion, mainly in the electric arc furnace;

- primary material ore - fusion reduction and refinement with use of oxygen.

Everything points towards intensive exploitation of the potential of recycled scrap, with the integrated method progressively losing its relevance, and being gradually confined to two functions:

- manufacturing new metal (a significant part of commercial steel cannot be recycled, and so a new source of metal is required);

- producing pure metal for specialised industrial use (a new area ripe for development, but one which will never be more than 30 to 40% of steel production).

Figure 1

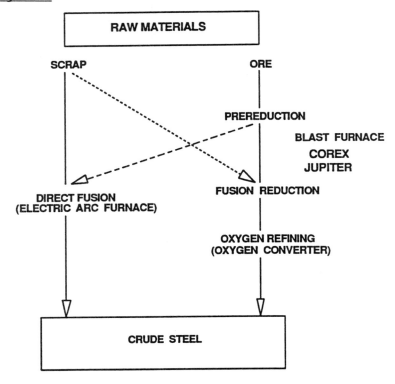

3   STEEL MILL OPERATIONS

As we've already seen, the furnace delivers a liquid
steel as far as possible without trace of slag, but
unfitted to be made from scrap (in the case of direct
fusion in the electric furnace) or from the liquid
casting (where there is indirect fusion from the
treatment of ore).

The simplest way of producing steel is to combine
the use of scrap as material with an apparatus as
technologically straightforward, yet as highly
productive, as the blast furnace. Such a combination,
depending on continuous flow and on simple lamination
techniques, is at the heart of the success of the
mini-factories with medium capacity (100s of 1000s of
tonnes p.a.) which have been started near great
industrial centres.

The electric furnace is not as old as the blast
furnace, but started around 1900 and has evolved to
become a mass producer of liquid steel. This increase
in power has impacted directly on furnace technology,
including operation at elevated pressure, and on

cooling techniques. Another development relies on new
technologies (the continuous flow furnace) and even
more on developments in furnace metallurgy. The most
obvious outcome is a reduction in consumption of
electricity from 500-600 to around 350 kWh. Pioneered
in Japan, this is now being pursued elsewhere.

A key factor among a whole variety involved in
these developments has been the new use of fossil
fuels, or more precisely coal injected into the bath
and then gasified into CO and CO_2. This does bring
some direct energy, but its main effect is to change
the properties of the dross through the release of
gas. This frothy dross tends to envelop the electric
arc, increasing its output, thus reducing heat loss
and electrical consumption. Using 10 to 20, or even
30, kilograms of coal per tonne of steel implies
considerable use of oxygen, not far short of that used
in oxygen processes. The oxygen acts both as a
combustible (by the oxidation on top of the bath from
CO to CO_2) and as a refining agent.

The future for the electric furnace seems clear.
The Japanese are already developing this new concept,
namely: a mono-electrode furnace using continuous
current; blowing inert gas in at the bottom to
homogenise the make-up of the liquid melt; wider and
better use of auxiliary combustible and of oxygen.
Developments are summarised in Figure 2.

Oxygen processes

The L-D (Linz-Donavitz) process led the field
when oxygen refinement was developed before the Second
World War. Its features are as follows:

(a)   Selective oxidisation of elements contained in
        the liquid. P goes into an oxidised liquid slag,
        with carbon leaving the molten mass as CO.
        What's left is a liquid mix of iron with small
        amounts of carbon and impurities, crude steel in
        fact.

(b)   10 to 20% of the CO is oxidised outside the
        molten mass into CO_2 (secondary combustion).

(c)   The operation is discontinuous but the process
        cycle is very fast, about 30 minutes from tap to
        tap in the best conditions.

(d)   Oxygen is blown in from the top onto the molten
        mass from a vertical cooled lance.

(e)   The liquid itself provides the energy (much of it
        contributed by the silicon oxidisation, the total
        oxidisation of the carbon into CO, followed by
        partial oxidisation into CO_2). This yields the

Figure 2

CLASSICAL PRIMARY METALLURGY

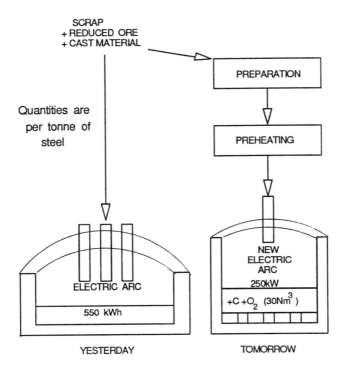

correct elevation in temperature for both steel
production and the formation of a liquid slag.
The excess of disposable energy allows quite a
lot of scrap to be fused (some 100-200 kilos per
tonne of hot metal).

(f)   Oxygen consumption is about 50-55 Nm^3 per tonne
      of hot metal under these conditions.

Over the last two decades the process has
evolved. The size of the converters has increased
from several tens of tonnes of metal to 200-300 tonnes
capacity. The increase in size of converters has not
added to the time taken by the process, considerably
increasing productivity. Yet there was a drawback.
The mixing of elements in the molten mass (made
possible by the release of CO gas) does not occur at
the end of the process; there is not enough carbon to
gasify. An iron oxide is then created, and this
degrades the steel quality.

To meet this problem a technique of blowing gas
in through the bottom of the converter has been

developed, either oxygen (the temperature has to be kept cool as it enters the bath) or an inert gas such as argon. Some processes push the gas injection at the base to the limits (e.g. OBM, LAS, etc.) but finishing with a light infusion of oxygen through the top to create conditions favourable to limited secondary combustion.

Today these original LD processes have been superseded either (a) by mixing top injection for most of the oxygen, with injection from the bottom for the argon and the remaining oxygen, or (b) by processes where the essential oxygen is injected from the bottom. <u>But whichever solution is chosen, oxygen consumption stays at 50-55 Nm^3 per tonne of hot metal</u>.

An alternative evolution has occurred in Japan. The Japanese have developed a technique of 'pre-refinement' in the last decade, which involves working on the crude liquid before it goes into the converter. They go further than desulphurisation, eliminating in advance of the converter all the silicon and most of the phosphorus. The converter then becomes merely a decarburator. But the standards have to be high as regards the final product, with strict control of temperature and of crude steel quality. This of course applies mainly to steel of great purity, which is increasingly the focus in Japan.

In Europe and the States pre-refinement has not been adopted for economic and other reasons, but I think myself it will have to be sooner or later. We will enter the era of ppm when discussing steel quality.

However, it does not much affect the consumption levels for oxygen. They are little affected as oxygen is mostly needed for decarburation. Oxidisation of silicon and the iron that goes into the slag clearly require much less oxygen. Some 45 kilos of carbon must be gasified per tonne of crude liquid to form a gas which consists of about 15% CO_2 and 85% CO.

It seems both reasonable and attractive to want to increase the amount of scrap melted in converters. (For steel manufacturers in the USA this has become a real obsession.) There are two ways of doing this which we reject: increasing the Si in the product, and oxidising the iron passing into the slag. Both have bad effects, with the latter damaging refractory linings and degrading the steel quality.

One solution is to boost secondary combustion, increasing the CO_2 leaving the reactor. The thermic output required for this is about 70%, quite reasonable. Increase the level of secondary combustion by 0.1 (say from 0.15 to 0.25) and the

scrap can be increased by 40 kilos, needing a further
$4Nm^3$ of oxygen per tonne. To do this the injection
method and the lance technology has to be altered - it
is feasible, though the results are limited.

To increase the level of secondary combustion a
supplementary source of energy is required such as the
addition of carbon. Coal infusion is the current
practice in the electric arc furnace. Conditions
would be even better in the converter as the reactor's
structure is better suited to transferring heat. 10kg
of carbon per tonne permits use of 60kg more scrap,
needing 12 Nm^3 more oxygen. In a certain number of
processes one can raise the injection level of coal
until there is no need to have recourse to external
melting (fonte extérieur). The possibility of
treating scrap in the converter is then considerable.
This is why a number of processes, first among them
KLOCKNER, have taken this route. At the limit,
without recourse to external melting up to one tonne
of steel can be produced from scrap - using 200kg of
coal and 200 Nm^3 of oxygen per tonne. So there is
scope for the traditional oxygen process to adapt to
new economic conditions by using more scrap.

However, the real strength of the converter will
be the processing of high purity steel (for the car
industry for instance) where scrap will never play a
large role. It is mainly a mass-production tool, but
finds itself part of an ever more fixed production
line. New developments such as direct rolling,
lamination and cooling imply the need for regularity
and continuity. However, the converter is badly
adapted to this function, particularly the melting of
solids. Overall it is unreliable, and it does not
meet the needs of modern steel-processing.

The future for steel-processing can be summarised
thus:

(a) For new metal: produce a primary carburated
     metal with equipment based on the blast-furnace
     or new techniques. The subsequent refining
     process will have no need for scrap. As the role
     of the converter will be confined to
     decarburation in the quest for a steel with 50
     ppm or less oxygen, it will be tempting to find a
     new technique for refinement comprising:

     -    a pre-treatment phase - removal of silicon,
          sulphur and phosphorus;

     -    one of progressively refining the carbon-
          traces down to around 1000 ppm;

     -    a final phase combining progressive
          decarburation with final purification with

respect to the traces of silicon, phosphorus and nitrogen.

But again, this won't have any great effect on oxygen consumption as such. Various process schemes are shown in Figure 3.

(b)   Steel from scrap based on the electric arc furnace. Fusion of scrap in the converter will be limited to a certain number of cases and can only be at the price of distancing the apparatus from its original structure. Though EOF, for one, has successfully used a variety of methods to achieve a high rate of combustion.

It is paradoxical that it is not in the basic oxygen process but in the electric arc furnace that new prospects for the oxygen industry are opening up. It is very different in the area of primary metal processing, or the 'fusion reduction phase' as we shall call it.

4   PRODUCTION OF CRUDE STEEL

We've seen that scrap is to be the main source for steel in the future. Yet some steel will still come from ore, both to offset losses from commercialised steel and to provide a metal pure enough for use in the steel production process itself. Tens of millions of tonnes in the EC alone are going to be lost from the traditional steel industry which is today guaranteed by the combination of blast furnace and oxygen process technology. We must take care not to be caught unawares by these developments as has often happened in the past. Two conditions must be satisfied in planning: a quantitative aspect to assure a level of production of commercial steel that is supposedly constant, and a qualitative aspect concerning the production of sufficiently pure steel.

To process metal from ore (based usually on $Fe_2 O_3$) we have to reduce, to get rid of, all oxygen present in the form of iron oxide. We also have to remove the matrix, which takes the form of a liquid iron-free slag with pronounced de-sulphurising properties.

As already said, pre-reduction is the term for reduction done in a solid state, and the term 'fusion-reduction' (*i.e.* reduction in the molten state) is appropriate as the rest of the process to eliminate oxygen is done in the liquid state.

Now let us see how these notions work out in practice.

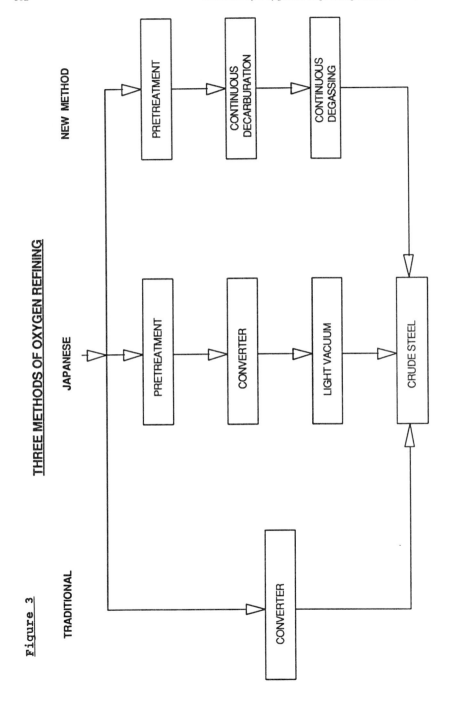

Figure 3

THREE METHODS OF OXYGEN REFINING

The blast furnace

The blast furnace has gone through a long process of technological adaptation and today brings the two principles just mentioned together in a single apparatus. Pre-reduction is done in the vat by a countercurrent exchange between a solid loaded on top, and the hot reducing gases emerging from the area of the tuyeres. The fusion part is done in the lower part of the apparatus. The fusion occurs in the molten state with the formation and decanting of the slag. It is accompanied by the residue of the reduction process. The subtle gasification of the carbon permits the harmonising of the energy balance and the oxygen balance.

After a pause in its metallurgic development in the critical 1970s, the blast furnace has a new lease of life from the injection of coal. In some cases this has exceeded 150 kilos to the tonne, with a correspondingly low consumption of coke. 300kg of coke per tonne is quite feasible today, where not so long ago 500kg would have seemed difficult. But so far the rate of super-oxygenisation of the air blown into the tuyeres has not increased in consequence. The blast furnace is primarily an exchanger in countercurrent mode and needs a minimal amount of gas to effect the heat transfer. The apparatus is so efficient in energy use, the volumetric rate of reacting gas (CO, CO_2, H_2 and H_2O) are insufficient to assure the thermic exchanges, and nitrogen must be added to assure the heat-carrying function. (See Table 2 for actual results achieved at Dunkirk No. 4. The second column gives estimates of what would be achieved with a maximum injection of oxygen.)

The lessons are these. First, that increasing the oxygen in the air blown into the tuyeres has hardly any effect on the thermic output of the equipment. Second, that the super-oxygenisation lessens the amount of air and gas needed (this could be helpful to an extent in eventually increasing hot metal production). Third, that super-oxygenisation, while it increases the adiabatic combustion temperature to the tuyeres, also reduces the temperature of the gas in the furnace's throat. As 100 degrees is the practical limit, you cannot push the rate of super-oxygenisation much above 25%.

So, if we keep to coal injection levels of 160-180kg, the steel producer won't have much interest in massive use of oxygen in the blast furnace and still less in building a central oxygen facility. However, if you lower the target for coke still further it could be a different matter. Before I get to that, a few words on new ways to increase the scope of oxygen use.

Table 2   Improved Blast Furnace

	INDUSTRIAL RESULTS	PREDICTED RESULTS
	Dunkirk No. 4	(calculated)
Temperature of gas °C	1200	1200
Coke kg/tonne	300	300
Coal kg/tonne	160	164
Oxygen Nm^3/tonne	20	60
Gas Nm^3/tonne	1000	830
% oxygen	22.3	25.8
Temperature at throat °C	170	100
Temperature of flame °C	2150	2270

Maximum oxygen consumption for 160kg of coal is $60Nm^3$/tonne.

New techniques

The Corex technique (discussed elsewhere by M
Schlebusch) was a daring invention whose success
inspires some respect. It differs from the blast
furnace in the way it integrates the solid state
reduction aspect. Not enough gases (CO, CO_2, H_2, H_2O)
to effect heat transfer are present. Also, nitrogen
must be kept out; it would cool the tuyeres, but raise
the temperature at the throat. So the Corex is quite
compatible with the use of oxygen in the order of 500-
$600Nm^3$ per tonne. The future of Corex will raise a
number of complex technical and economic issues which
we cannot address here. One essential problem is how
to utilise the considerable amounts of gas available
in the process.

The Jupiter is an EC version, based in France, of
a new technique being studied variously in Japan, the
United States, and Europe. It has two phases: pre-
reduction (solid state) in a fluidised bed; and final
fusion-reduction in a carburated molten state with
injection of the product from the first stage.

These are the great hopes at the moment, and both
put a lot of stress on oxygen - around $200Nm^3$ per

tonne for the fluidisation stage and 200-300 for the
hot liquid stage, totalling 400-500Nm^3 per tonne of
metal so processed.

The advantages of these two procedures over the
blast furnace are:

-    they are better-suited to medium-sized production
plans;

-    greater flexibility in the long run;

-    they bypass the ponderous preparations necessary
     for the blast furnace; the Corex makes coke ovens
     unnecessary, and Jupiter makes both coke ovens
     and ore sintering unnecessary. A comparison is
     given in Table 3.

Table 3   Two New Processes

NAME	STANDING	OXYGEN CONSUMPTION Nm^3 per tonne of hot metal
Corex	Industrial process	500 - 600
Jupiter	Process under study	400 - 500

Yet the blast furnace still has its place in the
sun, and this will perhaps continue for a while if it
continues to adapt as it has hitherto.

The new frontiers of the blast furnace

The way forward for the blast furnace may be
found in giving a further push to coal injection and
the reduction of coke consumption. Two factors which
will arise from increases in use of coal, and which
will sooner or later halt its progress, seem to me
vital:

The first concerns combustion, or rather
gasification of coal. In conditions such that
oxidising elements are hard to measure it is advisable
to avoid the accumulation of soots and gassy
concentrations which can't be later absorbed. Also to
restrict and control the formation of semi-coke from
the injected coal. Again Professor Prado's work is
relevant.

Secondly, as the quantity of coke reduces there is a significant growth in the proportion of carbon from coke gasified before it reaches the lower region of the furnace. It can reach and even exceed 50%. Then coke risks losing the necessary and sufficient cohesion to assure its place in the blast furnace.

How far can we go? Certainly down to 300kg, perhaps as far as 240kg. Experience will tell. There will be a need for oxygen anyway, first to maintain adequate levels of temperature at the tuyeres; heat released will tend to be reduced through coal injection. On the other hand, the use of superoxygenised air (the stock-in-trade of oxygen burners in particular) can only improve the prospects of coal gasification in the lower region of the furnace.

Table 4 gives results arising from the use of coal with about 30% volatile matter. Clearly the blast furnace could be a major user of oxygen, and the oxygen would not have to be all that pure. There are also even more imaginative ideas than coal injection, such as the elimination of the 'cowper' (heating stove or air heater) which is today both a sophisticated and an expensive piece of equipment. Two factors will diminish its role. Firstly, the more oxygen the less the volume of gas. Secondly, savings on coke from high gas temperatures are reaching their natural limits, leading some, such as A Poos, to suggest the direct injection of cold gas which opens up new vistas for oxygen use. I myself would go for a simplified continuous pre-heating of air, at heats of say 800-900 degrees.

Table 4   The New Conditions for the Blast Furnace

Gas temperature °C	1200
Coke kg/tonne	240
Coal kg/tonne	270
Oxygen Nm^3/tonne	180
Gas Nm^3/tonne	770
% oxygen	28.5
Temperature at throat °C	100

A second way forward looks at the exploitation of the tuyeres and the areas near them for the physico-chemical reactions of fusion-reduction. The presence of semi-coke in the product from the fluidised bed makes this material an ideal substance for injection at the tuyeres. Hot metal production could be boosted by up to 25% as the metal from the injection would escape in a countercurrent manner.

Regarding the source of supplementary energy, the ultimate in refinement would be to use hot plasma in the tuyere area, a technique which has been experimented with a little.

Figure 3 illustrates this new blast furnace idea, whilst Table 5 compares the results of:

(a) massive injection of semi-coke completed by use of natural gas

(b) more restricted injection semi-coke, use of natural gas, and plasma.

Table 5: New Blast Furnace: Comparison of Two Modes of Operation

	CASE A	CASE B
Gas temperature °C	900	900
% of melt pre-reduced	25	25
Coke kg/tonne	200	200
Semi-coke kg/tonne	150	90
CH_4 (Nm^3/tonne)	50	70
O_2 (Nm^3/tonne)	150	100
Gas (Nm^3)	400	440
Fluidisation stage:		
Coal kg/tonne	300	180
O2 (Nm3/tonne)	70	70

Both processes involve pre-reduction and the production of semi-coke in a fluidised bed. See text for details.

For the total output you have to allow for the pre-reduction, and also for coal used in the fluidiser. Two tonnes of coal is needed per tonne of semi-coke, and about $70Nm^3$ of oxygen per tonne of hot metal are consumed in this stage.

Round Table Discussion

Use of Oxygen in Iron and Steel Manufacture

Résumé of oral presentation of W. Schlebusch
DEUTSCHE VOEST ALPINE

The advantages of direct smelting were mentioned by M
Michard and Dr Corbett already; thus I will restrict
my explanations to our Corex plant in South Africa.
The research and development started at the end of the
1970s. This is mentioned in order to indicate how
long it took to get from an idea to a commercialised
plant. I think it will also be like this for other
concepts.

The plant in South Africa has an annual capacity
of 300,000 to 350,000 tonnes. It could also be a bit
more, depending on the ore feed. In South Africa at
the moment only lump ore is fed into the smelter
gasifier; therefore the production is limited to about
350,000 tonnes. The principle of the process has
already been discussed so only a few additional
remarks. The dust is separated in a cyclone and re-
injected into the smelter. This is an advantage
because the dust load of the gas is reduced. The
process has now been working for two years without any
major problems and we believe that this particular
arrangement is very important for the success of
Corex.

At the moment we are producing more than 43
tonnes per hour with 100% lump ore. The hot metal
analysis is similar to the blast furnace analysis, and
likewise the hot metal temperature. Slag rate is
quite high due to the coal quality. The coal quality
in South Africa is very poor and high in ash content.
On the other hand, this is a very cheap raw material.
Consumption rates for ore, coal and oxygen are high
per tonne of hot metal. The oxygen rate is
$540 Nm^3$/tonne hot metal. In common with other new
processes, this process consumes much more oxygen than
traditional processes.

The off-gas must be used in a valuable way. One
method is to produce electrical power. However, this
is not the only feasible way. In South Africa a

reduction unit for the export gas is used as a substitute for natural gas and coke oven gas. It is used in the plant for heating and reheating furnaces. This is the cheapest way from the investment point of view. The third way to use it is in a reduction unit for the production of 'sponge iron'. This is a very interesting process.

Electric arc furnaces also have good potential for O_2 consumption. Following research and development work on the injection of oxygen into the electric arc furnace, two KES systems are in operation in Italy. The increase of O_2 consumption from a high starting point has brought the O_2 per tonne almost to the level of a normal converter process. So even there there is a bright future for O_2 producers.

2   GENERAL DISCUSSION

Contribution 1 from the floor

Main problem for steel industry is the size of the required investments. Can you imagine the development of a smelting reduction process which used existing blast furnaces as the main vessel? With reduced investment costs the efficiency need not be so high.

Response from Dr Schlebusch

The vessel costs are not dominant; the special valves, the special instrumentation and other peripheral equipment contribute significantly to the investment. However, if you go with the Corex or something similar, the existing peripheral equipment can be re-used. Nevertheless, the opportunities in this area are limited. Hoogovens and British Steel have terminated a project which involved the cutting in half of a blast furnace with the upper part as a smelter and the lower half as a smelter. Perhaps Dr Corbett could say more.

Response from Dr Corbett

The reason we did not proceed with that particular development is not that we lacked belief in its technical feasibility. The economics looked marginal and at the same time we were conceiving the CCF development upon which we decided to concentrate. It must be remembered that new processes which seek to compete with the blast furnace are in fact competing with a highly efficient process unit. The off-gases are low in temperature ($100-120°C$) and so lean that they can hardly burn. This is the only energy the unit rejects. Thus, returning to the point about investment costs, new processes must be as efficient

as the blast furnace and so capital cost reductions
which reduce efficiency are unlikely to yield
competitive production costs. Needless to say, the
present margins are small and iron and steel companies
are struggling.

Contribution 2 from Professor Michard

I think this is an ancient dream which has done a
lot of harm, this mythical notion that direct
reduction can be used to transform ore into steel with
great speed.

When I joined IRCID in 1953 it was to work on
direct reduction, yet here I am forty years on talking
about the blast furnace. The truth is that metal-
processing combines a variety of unit operations, each
of which has its own part to play in achieving the
overall result.

These unit operations didn't all get built
through the operation of blind chance. Something like
a common purpose informs the various elements in the
process, and though they may seem more or less
randomly distributed, they are in fact quite close to
one another in terms of the contribution each makes to
a common result. There is, as you are aware, a vast
technological difference between the Corex and the
blast furnace, but there is no metallurgic difference
as such.

There is a logic, a sense of purpose almost,
behind all this. We are an old industry, experienced
enough in achieving real tangible thermodynamic
results. And in thermodynamics, the straight line is
not always the best one to follow ....

Contribution 3 from the floor

I would like to hear a little more about the
converter cyclone process which we learnt of this
morning. What is its stage of development?

Response from ILVA (Italy)

The process is very similar to Corex but it has a
single reactor. In the upper area we introduce the
iron and the O_2 in a cyclone region. This is a very
difficult area to project because iron ore must be
produced in the 'melter' zone. This is the critical
area of the process. Another thing is the 'tuyere'
area is quite similar to a gasifier and we introduce
O_2 and coal and we have the 'de-gasification' of coal.
It is also possible to inject O_2 with lances. We have
studied placement and O_2 levels in order to maintain a
high temperature of about 1500°C in this area. This
permits the produced ore to melt and fall down into

the gasifier.

Contribution 4 from the floor

Is this just a concept or has it already been
tested? If that is the case, what kind of size was
the trial?

Response from ILVA

We are trying to build a converter of about 5
tonnes per hour of hot metal. This is in a plant
which will be built as soon as possible. If all the
trials are successful, we hope to produce a bigger
converter with a higher productivity, similar to Corex
or a blast furnace. Designs are being developed.

Contribution 5 from the floor

We try to increase the quantity of coal injection
in the blast furnace, and it seems to be possible at
least to operate in the range of 160 or 170kg per
hour. Do you see a limitation in increasing this
quantity, e.g. 200, 250 or 300? I'm talking of
technical limitation or can we extrapolate? With this
additional coal injection you have a need for O_2. Do
you go through enriching the air coming from a
'cowper' or do you think the injection should be pure
oxygen through lances?

Response from Dr Corbett

One important feature of our blast furnace O_2 use
is that the pressure and the purity of the O_2 is not
as critical as for steel making. I think there is a
general interest for iron and steel businesses to
save, if possible, their O_2 costs and to compromise on
pressure and purity. Purity is not a major issue if
you're only going to mix it with air. I think that in
the longer term more O_2 will be injected directly
rather than put into the blast before the stoves. The
reason for this is that I start to worry about using
highly enriched blast in stoves which have
carbon/steel shells, with some refractory lining. An
enriched mixture can cut a hole in anything made of
carbon and steel extremely quickly. I feel more
comfortable injecting the O_2 at the last possible
moment before it goes into the furnace. Also I
believe that the combustion efficiency is improved by
having a 'lance' or a burner where the coal and the O_2
are very intimately mixed. A colleague indicated that
he had 'swirl veins' on the tip of his lance. We use
a dual concentric lance in our oxy-coal development at
Cleveland Iron. We certainly believe that we have
better combustion efficiency by mixing the coal and O_2
rather than simply infiltrating the O_2 into the hot
blast.

Contribution 6:   Question from D Deloche, L'Air Liquide to Jean Michard

You believe there will be a reversal of the 60% from blast furnaces to, say, 35% from electric arc furnaces during the next ten years. Do you think that during this period of change there will be an opportunity for new processes like Corex?

Response

Ten years is too long a timescale; we should think in terms of 20 or 30 months. With regard to Corex, it is always necessary to have evolution in iron and steel plants.

Contribution 7 from Dr Schlebusch

May I make another contribution on future process; I think this is quite interesting. We are of course trying to sell Corex. The question is, who is our typical client? Are they operating an integrated steel plant or electric arc furnaces? Today's 'minis' and 'midis' require flexibility of input material and also output material. Our first Corex plant in South Africa does not have a converter process behind the Corex but an electric arc furnace (e.a.f.). This e.a.f. is fed by 50% hot metal and 50% scrap as scrap is a rare material in South Africa. In the US people have to think about another feed material for their e.a.f. because the quality they produce needs virgin iron. In Italy a very small e.a.f. fed with hot metal is producing 6000 tonnes per year.

There are a lot of new scenarios in the steel industry and there will not be just one process. I think there is a lot of flexibility. On the other hand, converter processes are not without any development. For example, KLOCKNER developed coal injection into the converter and this has many possibilities for development.

Contribution 8 from Dr Corbett

I think the other thing to bear in mind is there's nothing like a bit of competition for sharpening up blast furnace iron makers. I recall that in the early 1980s Corex was just appearing as a process and there was a lot of interest in coal-based iron making. Our technical director went round all the iron works managers. He told them to sharpen up, to prevent going out of business. Because of the perceived threat from the new processes, technical advances in existing processes have been made. The blast furnace has still got a lot of mileage left in it.
At this point the discussion was concluded.

Historical Session: From Pollution to Protection

Where Even the Birds Cough: The First British Cases of Large-scale Atmospheric Pollution by the Chemical Industry on Merseyside and Clydeside in the Early 19th Century

P. N. Reed

NATIONAL MUSEUMS AND GALLERIES ON MERSEYSIDE, 127 DALE STREET, LIVERPOOL L69 3LA, UK

1 INTRODUCTION

Liverpool and Glasgow, as the main towns of Merseyside and Clydeside respectively, saw dramatic change in the early part of the 19th century - an expanding population, more effective transport, more varied and larger manufactures (often dependent on each other), expansion of commercial enterprise, and development as major international ports (accompanied by the associated businesses of insurance and banking). The population was increasing to a point where available housing was inadequate and often of poor quality, at least for the vast majority. It was a time of environmental crisis where the quality of air and water was called into question and diseases such as typhus and cholera reached epidemic proportions with high mortality rates.

This period also saw production of chemicals change from a small-scale trade to an industrial enterprise. This was nowhere clearer than in the alkali trade with the increasing demands for glass, soap and textiles. Until the end of the 18th century soda had been produced from vegetable sources - mainly barilla and kelp. Barilla - the ash of a Mediterranean plant - contained 14 - 20 percent of carbonate of soda, while the ashes of kelp - the brown Fucus seaweed found in abundance around the west coast of Scotland - contained up to 6 percent. The upheaval in Europe and the Peninsular War brought the supply of barilla to a halt, and kelp was unable to satisfy the demand for soda.[1]

In France the situation was more acute. In 1775 the French Académie des Sciences offered a prize of 100,000 francs (at that time worth about £4,000) for an effective way of producing soda from common salt. Many prominent scientists of the day experimented with possible solutions to this problem, but Nicholas Leblanc, surgeon to Duc d'Orleans, was judged to have submitted the best method.[2] This involved the reaction of sulphuric acid on salt to produce saltcake (sodium sulphate) which was then heated with coke and limestone. The resulting black-ash was lixiviated with water to produce a solution of carbonate of soda. Besides the soda, however, various by-products were produced, notably hydrogen chloride gas (or muriatic acid gas as it was known in the alkali trade) during the first stage, while the second stage produced the evil-smelling "alkali waste" or *galligu*.[3]

By the early 19th century coal smoke was already a major environmental nuisance, but the muriatic acid gas from the Leblanc alkali works was to become an additional blight on air quality. For every ton of salt converted into soda, half a ton of this acid gas was produced.[4] This paper examines the problems faced by manufacturers, landowners , local inhabitants and municipal authorities in Merseyside and Clydeside over the disposal of this noxious gas.

2 THE LEBLANC PROCESS ON MERSEYSIDE AND CLYDESIDE

The Leblanc process was introduced to England by William Losh in 1814 when, following a visit to Paris, he began using it at his Walker-on-Tyne works. On Clydeside, Charles Tennant began using the process at his St. Rollox works in 1818, and James Muspratt used it at his Liverpool works on Merseyside from 1823. As the demand for soda rose steadily during the first half of the 19th century, so similar works began to spring up in these areas.[5]

Muspratt arrived in Liverpool from Dublin in 1822. It may well be that he saw Liverpool as a potentially important centre for soda with its local soapboiling trade and its close proximity to the Lancashire textile industry. But there were other strategic advantages, namely that during the 18th century a triangular trade developed between Liverpool (as a thriving and fast-developing international port), the coalfields of St. Helens and the saltfields of Cheshire. The raw materials for the Leblanc process were therefore within a 30-40 mile radius of Liverpool and as the transport system improved they could be moved more efficiently within this triangle in larger quantities.[6]

In the late 18th century Scotland (and in particular the Clydeside region) was also a key area in the growth of the alkali trade. The vitriol works of Roebuck and Garbett was erected in 1749 at Prestonpans and by the 1840s it was importing Cheshire rock-salt from Liverpool. Large-scale developments in the alkali trade began with Charles Tennant, who had worked as a weaver and bleacher, before building the St. Rollox works on the Monkland Canal for producing bleaching liquor in 1798.[7] In the following year Tennant was granted a patent for dry bleaching powder (although it was the original idea of Charles Macintosh) and the name Tennant and the St. Rollox works became synonymous with bleaching powder - increasing its production from 52 tons to 910 tons in the first 25 years of the 19th century.[8] Chlorine was made from salt, manganese ores and sulphuric acid before 1803 when Tennant began using a very modified form of Leblanc process; it not was until 1818 that the full process was employed. The location of the St. Rollox works in relation to the raw materials for the Leblanc process was interesting - coal was available locally from the Lanark pits, but the salt came from Cheshire and the limestone from Ireland.[9]

Both Muspratt and Tennant were major entrepreneurs in the chemical industry in the early part of the 19th century - developing new products and searching out new markets (often overseas), and using the profits to develop and expand their businesses. They were both very active in developing the North American alkali trade.[10] However, expansion of their soda production brought additional problems - one of the most important being the large tonnages of the muriatic acid gas. In 1836 William Gossage had perfected a tower (originally adapting a derelict windmill) by which almost 100 percent absorption was achieved. This relied on the large surface area of contact between the water - in the form of a film spread over various inert materials such as bracken, brick, coke - and the gas.[11] For Muspratt this was too simple!! He was convinced that substantial quantities of gas could only be absorbed by using a large bulk of water. His outburst *"..Sure, all the waters of Ballyshannon itself would not suffice to condense the acid I make!"*,[12] reflects his Irish background but more importantly his lack of understanding of the principle of Gossage's invention. The latter is perhaps not too surprising when one remembers that (unlike his sons who were to follow him into the chemical industry) he had received little formal education in chemistry.[13] Neither manufacturer (along with most other alkali manufacturers) at this time was minded to alleviate pollution by condensing the gas in water and diversifying into other useful chemicals. Today it may be difficult to appreciate why they failed to use the Gossage tower, though their defence was the alkali trade was so important to the prosperity of the two regions and to the nation that the resultant pollution was the price to be paid. We are now so much more environmentally conscious and the incentives are clearer - the muriatic acid gas could be used for other products (thereby increasing cost-effectiveness) and the fines imposed by a court would make erection of a Gossage tower a cheap option. However, until the increased demand for bleaching powder in the 1860s, even adoption of the Gossage tower was not the complete answer, for the resultant hydrochloric acid was run out into rivers and streams - thereby only turning one form of pollution into another.

For Muspratt and Tennant the only approach was to dilute the gas with air and disperse it over as great a distance as possible. Their way of achieving this was by using tall chimneys. The chimney at Muspratt's works was about 250 feet high, and such chimneys became landmarks (and eye-sores!).[14] Not all the chimneys were associated with alkali works but they became the standard means for disposing of gaseous pollutants, and their number became an indicator of industrial activity. In the Report from the Select Committee on Steam Engines (1819) which had been asked to report on coal smoke nuisance, *"...a building inspector testified that twenty years earlier one could see the ships in the harbour of Liverpool from a nearby hill, whereas in 1819 chimneys hid the view of the harbour although the number of houses had not increased"*.[15]

In Glasgow, the St. Rollox works was by 1840 recognised as probably the most extensive in Europe, covering over 5 hectares. The scale can be judged by the fact there were 100 furnaces, retorts and fire-flues, and the weekly consumption of coal was 600 tons.[16] The resulting coal-smoke and fumes (from the Leblanc process amongst others) were dispersed from a chimney which was the highest in Britain and remained one of the landmarks of Glasgow until the 1920s. This chimney was called "Tennant's Stalk" and the following description published in *The Mirror* in December 1841 gives a clear impression of this formidable structure.
"The chimney is founded upon a bed of solid sandstone rock, twenty feet below the surface of the ground. The diameter of the

*outer chimney is fifty feet at its foundation, forty feet diameter at
the surface of the ground, and will diminish......to a diameter of
fourteen feet six inches, when it will have attained an altitude of
from four hundred and twenty to four hundred and thirty feet. The
inner chimney is a cylinder of sixteen feet diameter, rising perpen-
dicularly to a height of two hundred and sixty feet. This inner
chimney is unconnected with the outer one, but comes very nearly
in contact at its termination, allowing only space for expansion
arising from the temperature. The flues from the various parts of
the extensive works are introduced into the inner chimney through
four circular apertures, each seven feet six inches in diameter..."*[17]
In 1853, 12,000 tons of soda ash and 7,000 tons of bleaching powder
were produced. 2,000 tons of suphuric acid were produced for sale,
though the total production was likely to exceed 16,000 tons. With
these levels of production it is not difficult to envisage the vast quantities of coal-smoke and assorted fumes released from the "Tennant
Stalk" and their effect on the environment.

When Charles Tennant commenced the St. Rollox works in
1798   the site was almost completely isolated, allowing plenty of
room for dumping waste products. He was able to avoid litigation for
pollution since the works' location north of Glasgow enabled the pre-
vailing south-west wind to carry coal-smoke and noxious vapours
away from existing populated areas. By 1861 the works had become
part of a teeming industrial estate. This was in marked contrast with
Muspratt's predicament in Liverpool where his works alongside the
Leeds Liverpool Canal was close to inhabited parts of the town and
included within an existing "industrial estate". St. Rollox has closer
parallels with the development of Widnes as a chemical town from a
rural hamlet (although much earlier and on a smaller scale).

3 COMPLAINTS OF NUISANCE AND LEGAL CHALLENGES

The use of chimneys to disperse the muriatic acid gas (or mist as it
became in atmospheric moisture) proved ineffectual having little
regard for basic meteorology and chemistry, and the "acid" rained
down on the surrounding inhabitants and vegetation with appalling
effect. As the quantities of gas released increased so the complaints
became more numerous and vociferous.

In Glasgow, Charles Tennant managed to avoid the worst of
these complaints and was able to continue his operations at St.
Rollox well into the second half of the 19th century. In Liverpool
Muspratt found the situation more threatening. The topography of
the land surrounding his works presented particular problems - the
most crucial being that the muriatic acid gas was frequently blown
onto the district of Everton standing about 225 feet above and three-
quarters of a mile from the Muspratt works, and therefore at about
the same height as the top of the chimney. The inhabitants of Everton
soon realised they were being subjected to a pungent and nauseous
gas and began to point an accusing finger at Muspratt.

Muspratt was not the first in Liverpool (nor was he the last) to
be charged with creating a nuisance, for in 1770 the copper works of
Charles Roe and Company had been removed by Liverpool

Corporation because of pollution from sulphur dioxide. In 1831 the inhabitants of Everton brought an action against Muspratt for the damage caused by his chemical works.[18] Muspratt's lawyers managed to have prosecution witnesses excluded from giving evidence on the grounds that the case was being heard in Liverpool and they lived outside the borough. Only witnesses friendly to Muspratt were allowed to testify: soapboilers, scientific experts and doctors, who even went so far as to claim that muriatic acid gas was beneficial to people because it destroyed disease. This surprising claim caused someone to announce in the *Liverpool Mercury* newspaper *"...the formation of a Muriatic Acid Gas Joint Stock Company, to supply the town by means of pipes, with this most valuable article".*[19] On this occasion Muspratt escaped lightly with a shilling fine.

Muspratt had been lucky but even he was perceptive enough to realise that the days of his Liverpool works were severely numbered. Before the courtcase began he was looking to relocate his chemical works and in 1828 formed a partnership with Josiah Gamble in St. Helens (some 15 miles from Liverpool) where his soda would be important for the glass industry.[20] Again, within a short time of commencing operations he received complaints, and by 1831 the partnership was dissolved and he removed his works to Newton (about 5 miles from St. Helens) where besides the Leblanc process, experiments took place on an early form of the ammonia-soda process and with regenerating sulphur from alkali waste.[21]

However, by 1838 continued accusations about the Liverpool works, which was still in operation, came to a head and in April that year Liverpool Corporation indicted James Muspratt for "creating a nuisance". The proceedings of the court case provide an important insight into the difficulties of obtaining a conviction for such an offence, and led eventually to changes in government policy and adoption of specific legislation.[22]

First, was the problem of attributing damage to one particular source. Muspratt's works was located near to the centre of town in an area of 100 factories, including 12 chemical works, 23 distilleries, 17 soaperies, 16 breweries, 7 limeworks, 17 foundries, 2 gas-works, tanyards, 3 sugar houses, 3 colour manufactures, 5 waterworks, 13 steam mills and mortar mills. To all intent and purposes, an early industrial estate from which a variety of smoke and vapour was released into the atmosphere.[23] How could there be any proof of responsibility of a particular works without analysing the chimney emission suspected of causing the damage? This may seem a obvious question to ask, but it remained unanswered even though the necessary analytical techniques were available.

The prosecution called 49 witnesses, including traders, gardeners and doctors. One witness claimed that the gas affected his wife and children so badly that they had to leave the town to improve their health. A gardener testified that pear, apple, sycamore and horse chestnut trees had been affected and in many cases killed. There were also humorous moments such as when a gardener asserted that his cauliflowers had become more and more puny in size year by year, to which the defence counsel replied flippantly that if the works went on much longer the cauliflowers would be reduced to the size of peas! The defence called similar witnesses, one of whom from Everton asserted that his pear trees facing the Muspratt works did very well, while those that faced the other way did less well, and accounted for this by the fact that the east winds were very damaging for any form of vegetation.

There was no disputing that damage had been caused, but rather what and who were responsible. The prosecution and defence produced witnesses whose evidence relied on mere observation of cause and effect without any facts. Rather than providing clarification, they only caused further confusion and misunderstanding. Whether muriatic acid gas was responsible for the damage cited also raised the question as to how harmful (or beneficial) the gas was. While it was generally accepted that the gas was harmful, supporters of Muspratt displayed banners in prominent places playing down these effects, but unfortunately some became over enthusiastic by promoting beneficial properties - (reminiscent of the 1831 trial). The accusing finger was pointed at Muspratt since his works was a recent addition to the area (no damage of this magnitude had been evident before 1822), it was large in comparison with others as reflected in the height of the chimney and, finally, gas clouds were supposedly seen coming from his chimney.

A second difficult area for the judge and jury was the evidence provided by "scientific experts". Evidence given by such experts for the prosecution and for the defence were in direct contradiction, and appraisal of the evidence necessitated detailed scientific knowledge which neither the judge, the jury nor the two principal lawyers possessed. In the Muspratt trial, such evidence was given for the prosecution by a surgeon and a physician and for the defence by Professor Thomas Thomson, Professor of Chemistry in the University of Glasgow.

The judge found great difficulty summing up the proceedings and decided to separate out the commonsense evidence from the chemical evidence. With the former, the judge was adamant that prosperity of the alkali trade to Liverpool and to the country was not a valid reason for dismissing the case and allowing alkali works to continue causing damage. The chemical evidence proved more difficult. The judge admitted he knew nothing of chemistry and was therefore unable to offer any analysis of the relative merits of the evidence put forward by the two parties. He warned the jury: *"if you have any knowledge of chemistry, I would not advise you to place any reliance upon it, as a smattering in chemistry is exceedingly dangerous and it might give you more reliance upon it than you ought to have.......try, if you can, to make your mind as to it, but upon strict principle, and if any of you have made chemistry an amusement, I would not advise you to trust it"*.[24] The judge decided, effectively, that since the scientific evidence for both prosecution and defence could not be understood fully it was best dismissed from further consideration. Not very good guidance I would suggest in a case of this nature. Nevertheless, the jury returned a verdict of guilty.

This verdict brought to an end any ideas James Muspratt (and his sons) had of expanding the Liverpool works, indeed the works were gradually allowed to run down. But the 1838 trial was not the last. Muspratt's works at Newton was accused of caused damage by local farmers and landowners including Sir John Gerard, and ended in further litigation. Once several prosecutions had been secured the number of claims rose dramatically though often for relatively small amounts. There is evidence that some of the damage cited in these claims was fictitious but nevertheless the cases were still won. The Newton works was now viewed as being guilty of "causing a nuisance" by its very existence rather than as a result of operational misman-agement on limited occasions. Following another major action in the Liverpool courts in 1846 Muspratt spent about £6,000 constructing Gossage towers.....

> *"There were five towers filled with coke, and about four con-densers to each tower, and a stream of water was kept running constantly day and night.....They sent from their furnaces, in the course of a week, 77 tons of dry muriatic acid gas, which was completely condensed and for the purpose of condensation 1000 tons of water were used weekly......They had tested the perfectly complete system of the condensation apparatus used, by sending, upon two occasions, one of their men into the flues, and who received no injury by going there."[25]*

Presumably this was an early example of quality control!!

By 1851, however, Muspratt had decided to abandon Newton and transfer his chemical manufacturing activities to Widnes on the banks of the River Mersey. Widnes had changed rapidly from the rural hamlet in 1847 when John Hutchinson purchased land along-side the Sankey Canal for a chemical works, developing into a chemi-cal town by the 1870s, and is still a centre for ICI to the present day. The Muspratts empire included an extensive alkali works close to the North Wales coast near Flint, where they were able to escape the legal challenges faced by their earlier works.

4 CONCLUSION

The introduction of the Leblanc process to Britain as the first phase of the development of the heavy chemical industry satisfied the increasing demand for soda, but it also brought the problem of muri-atic acid gas. Before increased demand for bleaching powder and other chlorine-derivative compounds from the mid-1860s made its conversion into chlorine essential, the acid gas exacebated the already poor environmental conditions found in the towns of Merseyside and Clydeside (in particular Liverpool and Glasgow). With no incentive to use the gas, even though the Gossage tower had been invented in 1836, the only approach for manufacturers such as Muspratt and Tennant was disposal - in other words dispersal from tall chimneys. This proved an unsatisfactory solution, but the legal process to force change found difficulty in attributing responsibility and admitting evidence from "scientific experts". The laissez-faire phi-losophy of government left the situation unchecked until the 1860s when the government was forced to take notice under the weight of opinion from landowners, farmers, doctors, and the public. What was needed was an interventionist policy on the part of government and adoption of legislation that set some limit of emission of the gas from alkali works and included inspection to ensure such limits were met. This was only achieved with the Alkali Works Act of 1863 and subse-quent legislation.

REFERENCES

1.   A. and N. Clow, "The Chemical Revolution",Books for Libraries
     Press, New York, 1970, p.80.
2.   J.R.Partington, "A History of Chemistry", Macmillan, London,
     1962, Vol.3, p.563.
3.   W.A.Campbell, "The Chemical Industry", Longman, London,
     1971, p.26.
4.   W.A.Campbell, "The Chemical Industry", Longman, London,
     1971, p.36.
5.   D.W.F.Hardie and J. Davidson Pratt, "A History of the Modern
     British Chemical Industry", Pergamon, Oxford, 1969, p.25.
6.   T.C.Barker, "Lancashire Coal, Cheshire Salt and the Rise of
     Liverpool", Transactions of the Historic Society of Lanca-
     shire and Cheshire,1951, p.83.
7.   Charles Tennant (1768-1838), "Dictionary of National Biogra-
     phy", p.60.
8.   James Mactear, "On the Growth of the Alkali and Bleaching
     Powder Manufacture in the Glasgow District", Chemical News,
     1877, 35, p.23.
9.   John R. Hume, "The St. Rollox Chemical Works 1799-1964",
     Industrial Archaeology, 1966, 3, p.188.
10. G.W.Roderick and M.D.Stephens, "The Muspratts of Liverpool",
     Annals of Science, 1972, 29, p.294.
11. P.N.Reed, Entry for William Gossage, "Dictionary of Business
     Biography", Butterworths, London, 1984, Vol.2, p.616.
12. J.Fenwick Allen, "Some Founders of the Chemical Industry",
     Sherratt and Hughes, London, 1906, p.88.
13. J.Fenwick Allen, "Some Founders of the Chemical Industry",
     Sherratt and Hughes, London, 1906, p.70-71.
14. G.W.Roderick and M.D.Stephens, "The Muspratts of Liverpool",
     Annals of Science, 1972, 29, p.290.
15. Ann Beck, "Some Aspects of the History of Anti-Pollution
     Legislation in England", Journal of the History of Medicine,
     1959, 14, p.478.
16. E.W.D.Tennant, "The Early History of the St. Rollox Chemical
     Works", Chemistry and Industry, 1947, 66, p.671.
17. The Mirror, 4 December 1841.
18. G.W.Roderick and M.D.Stephens, "The Muspratts of Liverpool",
     Annals of Science, 1972, 29, p.291.
19. The Liverpool Mercury, 13 May 1831.
20. G.W.Roderick and M.D.Stephens, "The Muspratts of Liverpool",
     Annals of Science, 1972, 29, p.290.
21. D.W.F.Hardie, "A History of the Chemical Industry in Widnes",
     I.C.I., Birmingham, 1950, p.56.
22. "An Indictment of the Corporation of Liverpool against James
     Muspratt, 1838, before Sir John Taylor Coleridge and a
     Special Jury for a nuisance, alleged to proceed from his
     chemical works in Vauxhall Road", Papers in the Harold Cohen
     Library, University of Liverpool.
23. G.W.Roderick and M.D.Stephens, "Profits and Pollution: Some
     Problems Facing the Chemical Industry in the Nineteenth
     Century. The Corporation of Liverpool versus James Muspratt,
     Alkali Manufacturer, 1838, Industrial Archaeology, 1974,
     11,p.37.
24. "An Indictment of the Corporation of Liverpool........",
     Papers in the Harold Cohen Library, University of Liverpool.
25. Report of County Court case against Mesrs. Muspratt and Co.,
     The Liverpool Journal, 22 September 1849.

Removing Mountains: Industrial Waste and the Environment

N. G. Coley

DEPARTMENT OF HISTORY OF SCIENCE, FACULTY OF ARTS, THE OPEN
UNIVERSITY, MILTON KEYNES MK7 6AA, UK

Industrial waste is generally liquid effluent as distinct from gases,
fumes and vapours or solid refuse, but increasingly the term is used
to describe any waste product of a manufacturing process. Since the
late nineteenth century the quantity of industrial waste has greatly
increased; its treatment and disposal is now a branch of advanced
technology.

To use materials or energy wastefully squanders valuable resources
and should be avoided, yet waste is an inevitable by-product of
industry. The law of conservation of matter dictates that in any
chemical change a given mass of reactants is converted into precisely
the same mass of products. Chemical industry is therefore a
recycling operation in which materials are converted from one form to
another. Naturally the chemist aims to maximise the proportion of
useful products and minimise waste, but it is rarely possible to
eliminate waste altogether. During the past century the scale and
complexity of the international chemical industry has gown to such a
degree that the rate at which raw materials are converted into waste
begins to interfere with the natural cycles in the atmosphere, rivers
and oceans and on land.

Solid industrial waste comes from many different sources. Large
quantities of spoil are produced in mining operations and further
large quantities are produced by the extraction of metals and
minerals from poor grade ores. In the treatment of natural products
the separation of cellulose from wood for rayon or paper manufacture,
or glue and gelatin from bones, are examples in which organic matter
is extracted from large quantities of raw material. Waste is also
produced in processes where, although it might be possible in theory
to achieve 100% conversion, it makes better economic sense to accept
a lower yield. This applies to the manufacture of textiles, leather
and paper; in food processing and so on. Waste is also produced when
material used for a limited purpose is thrown away, for example
packaging. Finally, materials such as catalysts used in industrial
processes become spent due to mechanical damage, chemical
decomposition or dilution.

Table 1 shows that industrial waste forms only a small part of the
total quantity of waste from all sources[1]. Much of the waste from

Household refuse	90 million tons p.a.		
Waste from manufacturing industry (metal, paper, rubber, plastics, textiles, glass etc.)...	80	"	"	"
Waste from chemical industry ...	45	"	"	"
Agricultural waste...	1000	"	"	"
Food waste...	100	"	"	"
Extractive industries	300	"	"	"

Table 1.
(From: R.E.Hester(ed), Industry and the Environment, 1983, p.146)

manufacturing industry is technically recoverable though the necessary processes may be economically prohibitive. Chemical waste poses other problems, due to its complexity and the fact that it is often highly toxic.

From the beginning of the industrial revolution in the 18th century solid waste from chemical processes was simply dumped near the works, but by the mid-nineteenth century it was becoming clear that these noxious heaps would have to be removed.

Solid industrial waste is difficult to handle and one way of avoiding mountains of it is to discharge the material, mixed with large quantities of water, into rivers or the sea. In the manufacture of sodium carbonate by the ammonia-soda process for example, the waste product, calcium chloride, is formed in quantities approximately equal, weight for weight, to the alkali produced.[2] As the annual world production of sodium carbonate is of the order of 20-30 million tons, there is a large quantity of calcium chloride. The demand for it is small and most is discarded, but as it is produced in aqueous solution, a mountain of solid waste is avoided. The liquor is first allowed to settle and solid residues are removed to be placed in underground cavities in the salt measures caused by the extraction of brine. The remaining liquid effluent is then discharged.[3]

In many cases however, solid industrial waste causes difficult problems. First there is environmental pollution and the question of whether or not the waste can be disposed of without treatment, or must first be made safe. Then there are the economic issues. Solid chemical waste usually contains valuable materials as well as some of the energy used int he manufacturing process. It is desirable, if possible, to extract and re-use these valuable components. Suitable waste may be incinerated[4], although heat recovery from this process is not very common as the cost of the necessary equipment can rarely be justified. The re-cycling of metals, rubber, cellulose, paper, plastics and glass also poses economic problems and it is often cheaper to manufacture these materials afresh than to recover them from waste.

An early example of efforts to recover a valuable raw material from chemical waste concerns the sulphur content of alkali waste from the old Leblanc process for making sodium carbonate. Consisting mainly of calcium sulphide, alkali waste was stored in vast heaps near the alkali works. Evil-smelling and poisonous, it was slowly oxidised and hydrolysed in the air with the evolution of hydrogen sulphide and there was a mountain of it, since for every ton of sodium carbonate there was also a ton of this solid waste. In areas like Tyneside in the North-east of England these heaps of alkali waste, or 'galigu', grew - a dangerous nuisance and an eyesore. But besides the environmental pollution the alkali waste locked up sulphur from the sulphuric acid used in the initial stages of the Leblanc process. This represented a considerable financial loss and efforts were made to recover this sulphur.

The Chance-Claus process, established about 1888, was the only successful method to be used on a large scale. In this the alkali waste was first suspended in water and gases from lime-kilns, containing mainly carbon dioxide, were passed into it in excess to release the sulphur as hydrogen sulphide which was then oxidised to sulphur by burning in air over a ferric oxide catalyst. A simple summary of the reactions is shown in Table 2.

The sulphur sublimed in a very pure state and could be re-used, for example in sulphuric acid manufacture.[5] It will be observed however, that the solid waste was not **removed** by this process, but was merely converted into the more environmentally 'friendly' calcium carbonate. The 'mountain' continued to grow even after most of the sulphur had been extracted.

Coal tar, a waste product of coal-gas manufacture produced in vast quantities in the 19th century was transformed after about 1870 from an environmental nuisance into a raw material of great economic value by demand for the products of coal-tar distillation with the rise of the organic chemicals industry. Benzene, toluene, aniline, phenol, naphtha and anthracene formed the main feedstocks for the manufacture of organic chemicals and were in considerable demand until the rise of the petrochemicals industry at the end of the second world war.

In many cases however, the recovery of useful components from industrial waste has proved uneconomic. The manufacture of titanium dioxide pigments used in plastics, rubber and paint, gives rise to vast quantities of solid industrial waste. The commonest ore of titanium, ilmenite, a form of ferrous titanate ($FeTiO_3$), contains on average about 20% or less of titanium dioxide. Rutile, natural titanium dioxide, also occurs in smaller quantities. The main extraction process involves heating ilmenite ore with concentrated sulphuric acid to convert it into ferrous and titanium sulphates.

$$2H_2O + CaS + CO_2 = CaSO_4 + 2H_2S$$
$$CaS + H_2S = Ca(SH)_2$$
$$Ca(SH)_2 + CO_2 + H_2O = CaCO_3 + 2H_2S$$
The resulting gas was then burned in air in a
Claus kiln with a ferric oxide catalyst.

$$2H_2S + O_2 = 2H_2O + 2S$$

Table 2

Other methods use chlorine or slag from iron works. The details need not concern us, but the likely quantities of waste products are almost seven times the weight of titanium dioxide produced. Since the annual output of titanium dioxide is of the order of 2 to 3 million tons, the total quantity of waste is very great.[6]

In theory it would be possible to recover the sulphur or convert the waste into saleable products, but the processes are uneconomic. The strong sulphuric acid is rejected from the process at 10% to 20% strength. For re-use it would need to be concentrated to about 96% requiring large amounts of heat and the ratio of energy costs to the value of the sulphur extracted makes recovery uneconomic. During the evaporation slurries containing metallic salts are formed which are difficult to separate from the liquid and the salts remaining in solution contaminate the acid, reducing its value. Roasting the waste metallic salts is also a high energy process which like the evaporation process gives off sulphur dioxide, some of which may escape to cause atmospheric pollution.[7]

Alternative methods of using the waste have been introduced locally on a small scale with some commercial success since 1976. The sulphuric acid may be converted into gypsum, copperas, or ammonium sulphate. Other methods of extracting titanium dioxide have been employed using rutile or enriched ilmenite ores, but none of these methods has yet found wide application in Europe. Instead, most titanium dioxide plants have been sited near to fast-flowing rivers or the sea, so that the waste can be discharged by pipeline or by dumping from barges. This seems to have had little impact on the environment up to now, but these methods may not continue to be acceptable in the future as legislative controls on waste disposal are tightened.

Recent experience has shown that solid industrial waste may sometimes be mixed with domestic refuse and disposed of by techniques such as composting and sanitary land fill, with no serious risk to the environment.[8] Certain kinds of organic waste can be converted into usable compost by biological oxidation under controlled conditions. This involves a suitable microbial content and a nutritionally balanced substrate. The rate of composting depends on controlling the degree of bacterial activity by temperature, pH and the degree of aeration. The waste is ground to improve aeration and exposed to atmospheric oxidation for about three weeks. Organic waste usually contains a sufficient indigenous population of microbes but the ratio of carbon to nitrogen is the critical factor. If this is greater than 25 to 1 the process becomes slower; at less than 16 to 1 the nitrogen is lost as ammonia. When the nitrogen content is low the refuse may be mixed with nitrogen-rich waste like manure or digested sewage sludge. Aerobic conditions must be maintained as lack of atmospheric oxygen leads to problems including odour. The compost can be used as a soil conditioner, but is not successful as a fertiliser unless it is initially rich in nitrogen. For this reason there is no ready market for the product and the process has not so far been widely used.

Up to the 1970s it was thought that sanitary land fill using chemical waste would have to be restricted to the most inert substances only,

For a plant producing 50,000 tons of TiO_2 per annum:

Waste Product	Tons per annum
Copperas ($FeSO_4.7H_2O$)	140,000
Strong acid (as 100% H_2SO_4)	75,000
Weak acid (as 100% H_2SO_4)	40,000
Soluble $FeSO_4.7H_2O$	83,000
total	338,000

Table 3
(R.Thompson(ed) The Modern Inorganic Chemicals Industry,
RSC, 1978, p.369)

but as industry has been able to provide data to back claims that such methods, even using some forms of toxic waste, would not be an environmental hazard, land fill has received strong support in Europe and this is now the most widely used technique for removing mountains of industrial waste. However, it cannot be used everywhere. Certain geological formations where there are faults or an unsuitable arrangement of strata may prevent its use due to the risk of leaching of toxic substances into water-courses.

This introduces the need to transport the waste to another area where conditions are more favourable. Now transport is costly and it also introduces a weak link into disposal procedures, by raising the possibility of accidents and the spillage of waste in unsuitable areas. Yet, the transport of industrial waste is sometimes essential. Besides the need to find suitable land fill sites there are other cases in which the waste must be taken to a central treatment plant. The most dangerous waste is often very persistent and remains a hazard for a long time. These cases include highly poisonous waste such as cyanide, dioxin, mercury and other heavy metal compounds as well as radioactive waste from nuclear reactors.

Since the second world war radio-active waste created by the nuclear industry has caused big problems for chemists, environmentalists and Governments. By comparison with the quantities of waste produced by other industries, the amounts of nuclear waste are relatively small, but the technical problems of safe collection, treatment and disposal are daunting.[9] The long-term effects on the environment and the danger to human life are still uncertain and consequently the removal of these particular mountains is an especially sensitive environmental issue with social and political repercussions.

The two naturally occurring actinide elements used in nuclear reactors are uranium and thorium - plutonium, the other fissionable material is a product of the nuclear transmutation of these two elements brought about in the nuclear reactor itself. Uranium occurs widely in nature, some of the most important deposits being in the sandstones of Wyoming and Colorado, in the USA and the conglomerates of Witwatersrand in South Africa. Uranium ore usually contains only a few hundred parts per million of uranium and very large quantities of ore have to be treated in order to extract usable quantities of the metal.[10] This process itself produces large quantities of low

activity waste. Cast into rods the uranium is built into the nuclear
reactor where it undergoes changes resulting in over 30 fission
products, all of which are highly radioactive.

Plutonium is found in nature in only very small quantities due to the
short half-lives of most of its isotopes.[11] [239]Pu is by far the
commonest isotope produced in nuclear reactors and is used to enrich
the uranium fuel in certain types of reactor. It has a half-life of
24,400 years, long enough to permit handling in large quantities. In
the nuclear reactor the reactions are very complex, yielding a large
number of radioactive substances.[12] The chemical state of these
products, whether as elements or oxides, depends on the initial
oxygen to metal ratio in the fuel and the temperature gradient in the
reactor core.

Re-processing is the treatment of irradiated fuel to extract fissile
material (mainly uranium and plutonium) for re-use. Because of its
intense radioactivity the fuel must be handled by remote control
behind adequate shielding and precautions must also be taken to avoid
using at one time a mass of material sufficient to cause an
uncontrolled nuclear chain reaction - a criticality accident. The
most difficult stage in re-processing is the initial separation of
unwanted fission products from uranium and plutonium. The irradiated
material is first 'cooled' for 100 days to allow short-lived radio-
isotopes to decay.[13] The next stage of treatment uses nitric acid
which is relatively non-corrosive to the stainless steel vessels.
Actinide nitrates are readily extracted in many organic solvents but
the other fission products are not. The commonest extractant is tri-
n-butyl phosphate (TBP) in solution in an inert liquid hydrocarbon.

TBP forms addition compounds with uranium and plutonium nitrates and
by careful choice of a number of factors satisfactory separation of
the uranium and plutonium can be achieved. A large re-processing
plant can treat 7 metric tons of irradiated fuel per day.[14] After
extraction of the uranium and plutonium the remaining acid solution
is highly active waste (HAW). It contains about 99.5% of the non-
volatile fission products and higher actinides such as Np, Am, Cm
produced by neutron capture in the reactor. Only a small fraction of

Uranium Production

Country	Planned or produced (1975) (tonnes)	Planned or Actual production (c.1985) (tonnes)	Actual production
Argentina	50	(A) 504 (1983)	208 (1988)
Canada	3560	(A) 9244 (1984)	13233 (1988)
France	1700	(?) 3000	?
Gabon	800	(A) 1179 (1984)	710 (1990)
Niger	1200	(P) 5000	2270 (1990)
USA	9610	(A) 5769 (1984)	4080 (1990)
Others	-	(P) 6400	?

Table 4
(Extracted from Thompson, op.cit., p.424 and other sources)

the uranium and plutonium (0.5%) finds its way into the HAW. The waste is partly evaporated and stored in double walled stainless steel tanks shielded by several feet of concrete. In addition to the fission products the HAW contains other elements.[15]

These radioactive elements come from additions to the fuel, the cladding of the fuel rods, or the processing vessels. Due to the heavy metals it contains there is some sludge in the solution and it is stirred by currents of air. The waste emits appreciable quantities of heat (20 watts per litre after 1 year; 2.5 watts per litre after 10 years) and the storage tanks are fitted with internal cooling coils. An important drawback to tank storage is the degree of continuous surveillance required. Up to twenty or thirty years this may be satisfactory, but where really long half-life waste is involved other more permanent methods of disposal are needed. The variety of component elements makes vitrification in a glass or ceramic matrix the only viable method which seems certain to hold the HAW intact.

There are several methods of vitrification in use. In Britain the whole process is carried out in the final container. The liquid waste and a slurry of glass-making chemicals is mixed, evaporated, calcined and vitrified in a series of stages. The process has the merit of simplicity, there are no moving parts and the container is used for only one run, but it is not easy to optimise the conditions. In France a different procedure is used. The waste is first denitrated and calcined to form a glass frit. This is transferred to a melting furnace which is periodically emptied into storage containers. A similar process is used in Germany. In America several other methods have been investigated but the chief problem with all of them is to prevent the escape of volatile products and ensure that they do not generate large volumes of low activity waste. Once the radio-active waste has been vitrified the containers are stored in water-filled ponds or in an air cooled repository where they must remain for some years. After that, disposal in the deep ocean, under the ocean floor or in some geological formation which is likely to remain isolated from human contact for the foreseeable future has been proposed; secure containment for several hundred thousand years is considered necessary.[16]

While everyone will agree that mountains of solid waste are thoroughly undesirable, there is a stark contrast between the industrialists' view and that of the environmentalist. The latter

Fission products	... ...	39% by wt.
Al_2O_3	... ...	20 "
Fe_2O_3	... ...	11 "
Cr_2O_3	... ...	2 "
NiO	... ...	1 "
MgO	... ...	25 "
ZnO	... ...	2 "

Table 5
(Thompson, op.cit., p.456)

generally wishes to establish uniform standards of environmental control; the former recognises the need for a pragmatic approach. Industry must be profitable to survive. Cleaning up the environment costs money and can easily swallow up all the profits. It is then necessary to find a reasonable balance-point acceptable both to industry and to society as a whole. A wise industrialist will respect the wishes of society, but will also endeavour to educate the public on the issues involved so that standards of environmental care are set which do not damage the prosperity of industry while protecting the environment.

It is wrong to see the environmental debate in polarised terms with industry as the villain and the environmentalists as the good guys, although this is all too often how the debate is presented by the media. Cases become politicised and sensationalised. The dangers are exaggerated and situations distorted. If environmentalists and industrialists would co-operate, production methods which do not lead to mountains of solid waste could be adopted while new ways of removing waste already formed are developed. With environmental standards set at realistic levels this approach would ultimately solve the long-term problems and lead to a better rapport between industry and the environment.

References
1.   R.E. Hester(ed)., <u>Industry and the Environment in perspective</u>, Royal Society of Chemistry, London 1983, p.146.
2.   Suggested about 1810 by Fresnel, early forms of this process were worked briefly in Scotland about 1836 (L. Mond, <u>JSCI</u>, **4**, 527; Smith, iv., 1887, **6**, 699) and near Paris in 1855 (Schloesing and Rolland, <u>Ann</u>. <u>Chim</u>., 1868, **14**, 5). Ernest Solvay's process was first worked in France near Nancy in 1872 and in Britain by Brunner and Mond from 1874 at Winnington near Northwich, Cheshire, where it is still worked by ICI.
3.   R. Thompson (ed)., <u>The Modern Inorganic Chemicals Industry</u>, Royal Society of Chemistry, London 1977, p.129.
4.   Kirk-Othmer <u>Encyclopaedia of Chemical Technology</u>, 22 volumes, John Wiley, New York, 1970, vol.21 pp. 646-47; R.C. Corey, <u>Principles and Practice of Incineration</u>, John Wiley, New York, 1969.
5.   JSCI. (1888), **7**, p.162.
6.   Thompson, op.cit., p.369.
7.   Ib., p.371.
8.   Kirk-Othmer, op.cit., p.648.
9.   <u>Cleaning our Environment, A Chemical Perspective</u>, American Chemical Society, 2nd edn., Washington, D.C., 1978, p.407.
10.   Thompson, op.cit., p.425.
11.   Ib., p.427, Table 4.
12.   Ib., p.441, Fission products: mass number distribution.
13.   Ib., p.448, Table 8 shows the main fission products remaining in re-processed nuclear fuel after 100 days.
14.   Ib., p.454, schema showing the process used at Windscale, Cumbria in the 1970s.
15.   Ib., p.456. Components of HAW.
16.   A.S. Kubo and D.J. Rose, 'Disposal of Nuclear Wastes', <u>Science</u>, **182**, 1205 (1973); <u>ACS Report</u>, 1978, p.424.

An Inspector Calls: Legislation on Air Pollution in Britain and Its Chemical Implementation (Alkali Act, 1863, *etc.*)

L. Gittins
c/o DEPARTMENT OF ECONOMICS AND ECONOMIC HISTORY,
UNIVERSITY OF PORTSMOUTH, LOCKSWAY ROAD, MILTON,
PORTSMOUTH PO4 8JF, UK

When William of Normandy landed in England the chief fuel was wood. Air pollution in a wood-burning economy was not a serious problem because wood smoke is fairly benign, and wood-burning favoured dispersal of industrial activity rather than concentration. The use of coal changed this. Coal production began on a small scale at Edinburgh and Newcastle about 1200, and in other parts of Britain by 1300. The switch from wood was gradual and it was not until 1600 that coal was in general use.[1] At an early stage it was carried by sea from Newcastle, and eventually a fleet of several hundred ships served ports as far south as Plymouth. Coalfields, harbours, and river estuaries increasingly became areas of industrial activity.

Wood and coal yield mostly water and oxides of carbon when burned efficiently, but coal requires a higher temperature than wood for this to occur, and in addition bitumenous coal contains an average of 1.5 per cent sulphur. Air pollution by coal can be attributed chiefly to inefficient combustion and to sulphur dioxide. Coal smoke and fumes provoked numerous attempts to ban its use, but this was not possible because the advantages of coal were overwhelming. Wood was becoming scarce and expensive, while the supply of coal was almost unlimited and relatively stable in price.

Lime-burners were probably the first chemical manufacturers to use coal. Purchases of coal for use by lime-burners feature commonly in the accounts of castles. In 1278, for example, the lime kilns of the Tower of London took nearly 400 tons of Newcastle coal.[2] There were frequent complaints about smoke and fumes, and in 1307 lime-burners in London were forbidden to use coal.[3] About 1500 the salt boilers at Newcastle began using coal,[4] and saltmaking became a subsidiary of the north-east coal trade. One visitor commented in 1664 that salt boilers caused "such a smoke that one would think the town were on fire".[5] Copper and lead smelters, and alum and oil of vitriol manufacturers polluted the air with oxides of sulphur but these tended to be small-scale local nuisances. Until 1750 air pollution came almost entirely from the use of coal.

It was the development of an industrial process for making sulphuric acid that made large-scale production of chemicals possible, and gave rise to the serious pollution problems of the nineteenth century. Oil of vitriol used by the metal trades, and for

making nitric acid, was made by roasting iron or copper sulphates.
In 1736 Ward and White adopted a more efficient process when they
burned sulphur with saltpetre (potassium nitrate) and dissolved the
acid in water.[6] The disadvantages of this process were its slowness
and the high cost of the glass bottles in which the acid was
collected. It was not until 1746 when Roebuck and Garbett burned the
mixture in a lead chamber that large-scale production of sulphuric
acid became possible. Ward and White were said to have reduced the
price to one sixteenth of its former level to 18 to 30 pence per
lb.,[7] in 1774 Roebuck and Garbett claimed to have reduced the price
'by 30 per cent',[8] and by 1800 sulphuric acid was selling at about 7
pence per lb.[9] Cheap sulphuric acid was essential for the commercial
production of synthetic soda.

In 1737 Duhamel du Monceau demonstrated to the French Academy of
Sciences the possibility of producing soda from salt. Father
Malherbe went a step further in 1776 when he obtained soda by heating
sodium sulphate with charcoal and iron. Leblanc became interested in
alkali synthesis in 1784 and patented his process in 1791.[10] The
reaction was conducted in two stages. First, salt and sulphuric acid
were mixed and heated to yield hydrogen chloride and sodium sulphate
(salt cake). The sodium sulphate was then mixed with calcium
carbonate (limestone) and charcoal, and strongly heated to yield a
mixture of sodium carbonate and calcium sulphide (black ash).
Finally, the sodium carbonate was extracted from the black ash with
water to leave a residue containing calcium sulphide (alkali waste).
Leblanc's contribution was at the second stage where he used limestone to complete the reaction.

Stage 1 $NaCl + H_2SO_4 \longrightarrow NaHSO_4 + HCl$
              $NaCl + NaHSO_4 \longrightarrow Na_2SO_4 + HCl$

Stage 2 $Na_2SO_4 + CaCO_3 + 4C \longrightarrow Na_2CO_4 + CaS + 4CO.$

The commercial success of Leblanc's process required cheap salt
as well as cheap sulphuric acid. Unfortunately, salt was a
convenient item for the government in London to tax. In 1805 the
salt duty was raised to £30 per ton to help pay for the war against
France,[11] and early attempts to produce synthetic soda by Doubleday
and Easterby (1812) and Losh (1821) at Newcastle, by Lutwych and Hill
at Liverpool (1814), and by Tennant at Glasgow (1818) were hampered
to some extent by the price of salt. The duty was reduced to £4 a
ton in 1823[12] and abolished in 1825.[13] James Muspratt, one of the
first to take advantage of cheap salt, operated Leblanc's process in
a disused glass-works in Liverpool from 1823. It was a shrewd choice
of location. The works were by the Leeds-Liverpool Canal which
supplied water, served as a drain, and brought in coal, salt,
limestone, and sulphur, and gave easy access to the inland markets.
The largest users of soda were the makers of hard soap, and in 1824
8,400 tons of hard soap was produced in Liverpool alone.[14] This would
represent a market for more than 4000 tons of synthetic soda, if the
soap makers could be persuaded to use synthetic soda instead of kelp,
the vegetable alkali then used in Liverpool. This occurred by 1830
and is reflected in a rise in imports of sulphur from 5,600 tons in
1820 to 19,100 tons in 1833, and a decline in imports of barilla from
18,500 tons in 1823 to 9,200 tons in 1840.[15]

"The Leblanc process of the early days made one chemical and wasted two"[16] - the chlorine in the hydrochloric acid and the sulphur in the alkali waste. It was clear that the hydrochloric acid liberated in the making of salt cake would be troublesome. Leblanc dissolved the acid in water in stoneware jars, Hill and Lutwych condensed it in shallow tanks, and Tennant condensed it in earthenware condensers.[17] Muspratt appears to have let the acid escape. The gas destroyed vegetation in the Everton district of Liverpool, and in 1831 the inhabitants brought an action against him. In 1835 he built a 250ft chimney to disperse the gas over a wide area and minimise the damage, but the nuisance continued and in 1838 Liverpool Corporation won a court action against him.[18] About 1831 he opened a works at Newton near St. Helens where he was obliged to pay compensation in excess of £500 on a number of occasions. In 1846 one claim cost him £1000,[19] and in 1851 he lost a further action for damages. He commenced production at Widnes in 1852, and closed the Newton works in 1854. The works at Liverpool and Widnes were still producing alkali in 1868.[20]

The importance of access to raw materials and markets was not lost on other chemical manufacturers. By 1852 at least 80 per cent of the alkali industry labour force of some 6,000 men worked at Newcastle, Glasgow, and at St. Helens and Widnes.[21] In 1862 St. Helens was said to be subjected to smoke and fumes from seven or eight alkali works, six or eight large copper smelting works, and a large number of collieries, glass works, and other factories.[22] The House of Lords was persuaded by Lord Derby to appoint a Select Committee "to inquire into the Injury resulting from noxious Vapours evolved in certain manufacturing Processes, and into the State of the Law relating thereto".[23] The Committee heard evidence about pollution from alkali, sulphuric acid, and alum works, and from copper and lead smelting. It was considered that alkali and copper works were the principal cause of injury. Hydrochloric acid vapour from alkali works was sometimes "perceptible" five or six miles away, and its effects within a radius of one or two miles were "fearful".[24] The Committee was told of offensive vapours arising from "thousands of tons" of alkali waste accumulating in or about St. Helens,[25] which contained calcium sulphide, calcium hydroxide, and a considerable amount of coal residue. When exposed to acid gases, poisonous and bad smelling hydrogen sulphide was released into the air and calcium chloride seeped into the nearest river. About 1½ tons of residue were produced for every ton of alkali and its disposal presented a serious problem. One witness told of two acres of land being purchased at £1000 per acre for the sole purpose of depositing waste.[26] The Committee also heard that an estimated 36,000 tons of 'sulphuric acid' in coal smoke fell on St. Helens every year.[27]

A complainant had two legal remedies, an action for damages or an indictment for nuisance at common law. If he sought redress he had first to identify the source of the nuisance. This was not easy when a dozen or more works were emitting gases. Even when an action was successful the penalty or compensation awarded was often inadequate - in 1828 Muspratt was fined one shilling[28] - and seldom was the nuisance stopped. Where gases were considered to be a continuing public nuisance a local authority had power to bring an indictment, but proving a case was difficult and this procedure was rarely used. It was clear that the problem of chemical pollution was

beyond the scope of existing law. The Lords' Committee considered
the Acts of Parliament since 1848, relating to Town Improvement,
Public Health, Removal of Nuisance, and Smoke Prevention, which gave
a measure of legal protection against air pollution, but some applied
only to London, others were effective only where they had been
adopted by the local authority, and in some the language was vague.
The Committee reported in July 1862 and recommended that the laws
respecting nuisances should be consolidated and applied uniformly
throughout the country; that the Smoke Prevention Act, 1854 respecting offensive trades be made of universal application; and that
independent medical inspectors should be appointed to supervise all
works producing noxious vapours. [29]

The subsequent Alkali Act, 1863, required every alkali works
where hydrochloric acid was produced to be registered with an
Inspector of Alkali Works. Inspectors were empowered to visit these
works and ascertain that not less than 95 per cent of the gas evolved
was 'condensed' (dissolved in water). [30] The acid fumes from copper
smelters were acknowledge to be bad but, because the fumes were
believed to be too dilute to be profitably employed, [31] they were not
covered by the Act. Similarly, the Act did not deal with alkali
waste because no way was known of extracting the sulphur.

The first Inspector appointed was Dr R Angus Smith, a chemist.
He reported in 1865 that 64 works in the United Kingdom produced
173,000 tons of dry hydrochloric acid per annum, of which almost 99
per cent was condensed. [32] The Act was not particularly onerous, it
merely required alkali makers to use a known and proven method of
condensation. Gossage at Stoke Prior, Worcestershire, had previously
devised a tower in which hydrochloric acid was absorbed by water
trickling through coke. The first tower was erected about 1836 at
Messrs. Crosfields, St. Helens, [33] and by 1840 three had been erected
at St. Helens and three at Newcastle, all of which it was claimed
gave "very satisfactory" condensation. [34] Condensation was not always
employed before the Act because there was no financial incentive.
Gossage told the Lords' Committee that the demand for hydrogen
chloride to produce chlorine for bleaching powder or carbon dioxide
in the production of bicarbonate of soda, utilised only about a
quarter of the total produced. [35]

The effect of the Act was to divert into a stream much of the
acid that had previously been carried by the wind. At that time
recovering chlorine from hydrochloric acid was inefficient and
expensive because the manganese dioxide used for the reaction was
converted to manganese chloride and was lost with half the chlorine.
River pollution became so serious that a Royal Commission was
appointed, and sat in 1869 to inquire into the matter. [36] The
Commission heard that about 45 per cent of the 244,000 tons of dry
hydrochloric acid produced in 1867 was run to waste. [37] Pollution
from the Leblanc process was eased to some extent with the adoption
of processes developed by Weldon (1866–1869) and Deacon (1868–1876).
Weldon recovered manganese dioxide by mixing the manganese chloride
with an excess of lime and passing air through the mixture.

$$MnCl_2 + Ca(OH)_2 \longrightarrow Mn(OH)_2 + CaCl_2$$

$$Mn(OH)_2 + \tfrac{1}{2}O_2 \longrightarrow MnO_2 + H_2O$$

Deacon avoided using manganese dioxide altogether. He passed a
mixture of hydrochloric acid and air at high temperature over a
catalyst of copper chloride and obtained chlorine and water.

$$4HCl + O_2 \rightleftharpoons 2H_2O + 2Cl_2$$

It was about this time the demand for chlorine from the textile and
paper industries increased sharply. Following the removal of an
Excise duty on paper in 1861, the paper manufacturers turned their
attention to straw, esparto, and wood, which had to be bleached, as a
substitute for rags which were becoming scarce. Muspratt noted that
bleaching powder production in the United Kingdom between 1861 and
1885 rose from 20,000 to 132,000 tons, and added, "The problem before
the Leblanc maker [in 1886] is a very difficult one, for he not only
has to face the competition of the ammonia-soda maker in the price he
obtains for his alkali, but if he increases the quantity of bleaching
powder made and the consumption remains the same, the price of his
product must also fall".[38]

Muspratt was referring to Solvay's ammonia-soda process. This
reaction was first described by Vogel in Germany in 1822, but its
industrial application was delayed until 1873 by chemical engineering
problems. A concentrated salt solution was saturated with ammonia
gas and then mixed with carbon dioxide in a Solvay tower to
precipitate sodium hydrogen carbonate. The sodium hydrogen carbonate
was then separated and heated to convert it to sodium carbonate. The
carbon dioxide evolved was recovered and used again, and so was the
ammonia.

$$NaCl + NH_4OH + CO_2 \longrightarrow NaHCO_3 + NH_4Cl$$

$$2NaHCO_3 \longrightarrow Na_2CO_3 + H_2O + CO_2$$

The process did not require sulphuric acid; it produced neither
hydrochloric acid nor alkali waste; and it used about half the coal
needed for the Leblanc process.[39] It produced soda more cheaply than
Leblanc and, ironically, it was only the demand for bleaching powder
later that kept some Leblanc works going.

Sulphuric acid was inconvenient and dangerous to carry in bulk
and by 1850 it was usual for alkali works to operate their own lead
chambers. Sulphuric acid was made with sulphur, but supply
difficulties in the 1830s encouraged alkali makers to experiment with
pyrites (copper and iron sulphides). In 1839-1840 Muspratt burned
pyrites at Liverpool and Newton. There were some problems but these
were resolved, and by 1852 St. Helens and Widnes were working almost
entirely with pyrites. Sulphur produced from pyrites was cheaper
than imported sulphur, and between 1859 and 1869 imports of pyrites
into the Mersey increased from 26,000 to 133,000 tons. Alkali works
polluted the air to some degree with oxides of sulphur and nitrogen
escaping from the lead chamber. Economies were achieved in the 1870s
by the widespread adoption of Gay Lussac's process (1835) for
recovering oxides of nitrogen by absorbing them with concentrated
sulphuric acid. The process was operated in tandem with Glover's
process (1859) which subjected the 'nitrous vitriol' from the Gay
Lussac tower to hot gases from the pyrites roasters. This
concentrated the sulphuric acid further and recovered more oxides of

nitrogen. The gases were then passed back to the chamber. These
improvements in the sulphuric acid process, when combined with the
use of cheaper pyrites, reduced the cost of sulphuric acid raw
materials between 1861 and 1886 by about 50 per cent.40 These
economies improved the ability of the Leblanc process to compete, and
extended its life.

Gas escaping from sulphuric acid plant did not escape the notice
of the Alkali inspector, and the Alkali Amendment Act, 1874 widened
the definition of a noxious gas to include sulphuric acid, sulphurous
acid (except that arising from combustion of coal), nitric acid,
oxides of nitrogen, sulphuretted hydrogen, and chlorine. The owner
of every alkali works in addition to condensing hydrochloric acid had
to use "the best practicable means of preventing discharge into the
atmosphere of all other noxious gases arising from such work, or of
rendering them harmless when discharged".41 Chemical manufacturers
resented the extra costs this imposed upon them, and when it was
apparent that the 1874 Act was not working well a Royal Commission
was appointed "to inquire into the working and management of works
and manufactures from which sulphurous acid, sulphuretted hydrogen,
and ammoniacal and other vapours and gases are given off....".42 In
the light of its report in 1878 the Inspector was given more power in
1881 in controlling noxious gases from alkali works, and his powers
were widened to include the supervision of works producing sulphuric
acid, chemical manure, gas liquor, nitric acid, sulphate of ammonia,
and chlorine and bleaching powder.43 The Inspector's powers were
further widened in 1892 to cover a total of thirteen chemical
activities including extraction of zinc ore, tar distilling, fibre
separation where hydrochloric acid was used, and the recovery of
alkali waste.44

Initially there was no way of dealing with alkali waste except
to dump it. Numerous attempts had been made to solve the problem.
Gossage said he spent some £20,000 and thirty years of his life
without any beneficial result.45 In 1862 Mond patented a process in
which air was blown through the waste, then the waste was washed with
water, after which hydrochloric acid was added to precipitate the
sulphur. It was not an ideal solution, because less than half the
sulphur was recovered. By 1888 Chance and Claus had developed a
process where carbon dioxide was blown through alkali waste to
produce hydrogen sulphide. This was then passed through a kiln
containing ferric oxide catalyst to separate the sulphur. Up to 85
per cent of the sulphur was recovered. Whether recovery was
worthwhile depended on the cost of sulphur. Not all alkali
proprietors were willing to invest in plant and the waste heaps
continued to grow. In 1881 the government, prompted by the
Inspector, attempted to ameliorate the problem. Alkali makers were
required to keep acid and alkali waste apart; they had to use the
best practicable means of depositing alkali waste to prevent it
causing nuisance; and the Inspector was empowered to serve notice on
owners of existing waste heaps that caused nuisance.46 By 1891 500
acres of land at Widnes had been covered with 10 million tons of
alkali waste to an average depth of 12 feet.47 The Inspector in his
1892 report said he was insisting that alkali waste was deposited in
successive layers of moderate thickness, instead of in embankments of
20 to 30 feet high, so that it would consolidate and exclude air.48

The last Leblanc plant closed shortly after 1918, the process having been worked in Britain for almost 120 years. For much of that time there had been no profitable way of utilising all the hydrochloric acid and calcium sulphide produced. The resulting pollution was particularly devastating because the nature of the process obliged large-scale producers to establish their works in a restricted number of locations. The Leblanc process was a weight-losing process; at first some 7-8 tons of raw material, including about 5 tons of coal, were used to produce 1 ton of soda. The process was strongly anchored to a coalfield. In 1864 18 alkali manufacturers were located at or near Newcastle, 8 at St. Helens, 7 at Glasgow, and 6 at Widnes - nearly 50 per cent of the total number registered.[49] These manufacturers were probably producing 80 per cent or more of alkali made.[50] St. Helens and Widnes had easy access to coal from the St. Helens coalfield, and rock salt from Cheshire. Newcastle, similarly, had no problem obtaining coal but was at a disadvantage with regard to salt. By 1830 Newcastle was using Cheshire salt almost exclusively. In 1852 Newcastle alkali works used 58,000 tons of salt against Lancashire's 40,000 tons;[51] in 1864 there were more alkali works at Newcastle than at St Helens and Widnes together. Newcastle produced a substantial amount of hydrochloric acid and had serious pollution problems.[52] Why did the matter of pollution come to a head at St. Helens and not at Newcastle? The chief reason was that Lord Derby had a country estate four miles west of the alkali works at St. Helens. This was to have an important bearing on the manner in which chemical pollution was tackled in Britain.

The 14th Earl of Derby, three times Prime Minister, was an experienced politician. This is reflected in the skilful way he brought the problem to the attention of the House of Lords in 1862. While pointing out the grievous nature of the nuisance he emphasised that he did not wish "to imitate the principles or practice of the French law, the provisions of which were very stringent", he wished to avoid legislation which would "unduly hamper any of the manu-facturing interests of the country".[53] The Lords duly elected Lord Derby chairman of the Select Committee. Witnesses appearing before the Committee were carefully chosen, and the alkali makers, realising 'the game was up', admitted that there would be no problem condensing the vapours. The recommendations of the Lords' Committee were fairly wide-ranging but the Alkali Act, 1863 confined itself to one problem; it dealt only with escapes of hydrochloric acid. It was an effective Act because there was a known remedy which was easy to supervise, and it was the thin end of a wedge. In 1874, 1881, and 1892 condensation requirements were made more rigorous; the emissions of other chemical manufactures were brought within the scope of the law; and an attempt was made to reduce the offensiveness of alkali waste. These successive extensions of the law were very largely made on the recommendation of the Alkali Inspector in his annual reports, and it is to him that much of the credit is due for reducing chemical pollution of the atmosphere in nineteenth-century Britain.

Neither the law nor alkali makers' consciences appear to have had any influence in promoting chemical remedies for pollution. The Leblanc process produced hydrochloric acid and sodium carbonate in fixed proportions. Optimum profits would be achieved only if the demand for the products matched these proportions. At first the

supply of hydrochloric acid exceeded demand. Alkali makers remedied
this by releasing the surplus into the air or, after 1863, into a
river. Weldon and Deacon, Mond and Chance/Claus were inspired by the
possibility of making profit from a product which would otherwise go
to waste. Similarly, Solvay, Gay Lussac, and Glover devised their
processes not as remedies for pollution but to use resources more
efficiently. Once ammonia-soda came into production the only
advantage possessed by the Leblanc process was that it produced
hydrochloric acid from which chlorine could be obtained. This
advantage disappeared with the opening in 1894 at Widnes of an
electrolytic process which decomposed salt solution to give chlorine
and caustic soda. Pollution by the Leblanc process eventually came
to an end because it was no longer profitable to continue polluting.
The profit motive outweighed by far any desire to reduce pollution.

The long-standing problem of urban and industrial pollution by
coal smoke remained largely untouched. Newcastle had endured smoke
since salt was first boiled there. In Glasgow in 1840 it was
observed, "The air is uncommonly good, if you get out of the
smoke".[54] In 1892 the Inspector pointed out that Widnes and St.
Helens each consumed about a million tons of coal a year for
manufacturing purposes. He estimated that 12,000 tons of sulphur per
square mile were deposited on St. Helens, compared with 11 tons per
square mile on London.[55] Coal smoke was accepted as an inevitable
fact of industrial life. This was acknowledged in the 1874 Act, and
in subsequent Acts, where sulphurous acid was defined as a noxious
gas, except when it arose from the combustion of coal. Relative
freedom from smoke was not to be attained in Britain until the middle
of the twentieth century. The problem of sulphur emissions is with
us still.

REFERENCES

1. R.L. Galloway, 'A History of Coal Mining in Great Britain',
    Macmillan, London, 1882. David & Charles reprint 1969;
    J.U. Nef, 'The Rise of the British Coal Industry', (2 vols.),
    Routledge, London, 1932. Cass reprint 1966.
2. P. Brimblecombe, Industrial Air Pollution in Thirteenth-Century
    Britain, Weather, 1975, 30, 391.
3. Nef, op. cit., Vol. 1, p. 157.
4. P. Pilbin, A Geographical Analysis of the Sea-Salt Industry of
    North-East England, Scottish Geographical Magazine, 1935, 51, 22.
5. 'Victoria County History', Durham, 1907, 2, 298.
6. S. Parkes, 'Chemical Essays', London, 1823, 2nd. edn., Vol. 1,
    pp. 474-5.
7. ibid.; H.W. Dickinson, The History of Vitriol Making in England,
    Transactions, Newcomen Society, 1937-38, 18, 48.
8. House of Lords, Roebuck v. Stirling (1774), 'Brodix's Patent
    Cases', B.V. Abbott, ed., Washington U.S.A, 1887, Vol. 1, p, 12.
9. J. Mactear, History of the Technology of Sulphuric Acid,
    Proceedings, Philosophical Society of Glasgow, 1881, 13, 423.
10. J.R. Partington, 'A History of Chemistry', Macmillan, London,
    1962, Vol. 3, pp. 70, 562-4; J.G. Smith, 'The Origins and early
    development of the Heavy Chemical Industry in France', Clarendon
    Press, Oxford, 1979, p. 194.
11. Statute 45 George III c.14, s.1. (1805)
12. Statute 3 George IV c.82, s.1. (1822)

13. Statute 5 George IV c. 65, s. 1. (1824).
14. British Parliamentary Papers (B. P. P.) 1826/27, XVIII, Account of Soap made in each town of Great Britain.
15. B. P. P. 1824, XVII, Articles Imported into G. B. 1814-1823; B. P. P. 1842, XXXIX, Statement of Imports and Exports into U. K. 1831-1840.
16. D. W. F. Hardie, 'A History of the Chemical Industry in Widnes', Imperial Chemical Industries Ltd., England, 1950, p. 43.
17. J. G. Smith, op. cit., p. 217: B. P. P. 1865, XX, First Annual Report of Inspector of Alkali Works (1st Annual Report of Inspector of Alkali Works), p. 14; J. Mactear, On the Growth of the Alkali and Bleaching-Powder Manufacture of the Glasgow District, Chemical News, 1877, 36, p. 26.
18. G. W. Roderick and M. D. Stephens, Profits and Pollution: Some problems facing the chemical industry in the nineteenth century. The Corporation of Liverpool versus James Muspratt, Alkali Manufacturer, 1838', Industrial Archaeology, 1974, 11, 35-45; T. C. Barker and J. R. Harris, 'A Merseyside Town in the Industrial Revolution St. Helens 1750-1900', (Liverpool University Press, 1954), Reprinted Cass & Co. London, 1959, pp. 225-231.
19. B. P. P. 1862, (486), XIV, 1. Report from Select Committee on Injury from Noxious Vapours (S. C. Noxious Vapours), paragraphs 60, 92, 280-287, 939-940, 1304; A. E. Dingle, The Monster Nuisance of All: Landowners, Alkali Manufacturers, and Air Pollution, 1828-1864, Economic History Review, 1982, 529-548; Barker and Harris, op. cit., p. 239.
20. B. P. P. 1870, XL, Vol I [c. 377] & Vol. 2 [c. 109], 1st Report of the River Pollution Commission, 1868, Vol 2, p. 131, 133.
21. S. C. Noxious Vapours, para. 1507.
22. S. C. Noxious Vapours, para. 13.
23. Hansard, House of Lords, May 9, 1862, columns 1452-1467.
24. S. C. Noxious Vapours, para. 781.
25. S. C. Noxious Vapours, para. 243.
26. S. C. Noxious Vapours, para. 246.
27. S. C. Noxious Vapours, paras. 1811, 1820.
28. Hardie, op. cit., p. 18.
29. S. C. Noxious Vapours, Report, pp. VIII, IX.
30. Statute 26 & 27 Victoria c. 124. 'Alkali Act, 1863'.
31. S. C. Noxious Vapours, paras. 1101-1108.
32. 1st Annual Report of Inspector of Alkali Works, p. 6.
33. B. P. P. 1893-94 XVI, 29th Annual Report of Inspector of Alkali Works, p. 12.
34. 1st Annual Report of Inspector of Alkali Works, p. 8.
35. S. C. Noxious Vapours, para. 971.
36. B. P. P. 1870, [c. 37] & [c. 109], XL, 'First Report of the River Pollution Commission, 1868'.
37. 1st Report of the River Pollution Commission, Part IV, p. 291.
38. E. K. Muspratt, President's address on the History and Development of the Manufacture of Alkali..., Journal of the Society of Chemical Industry, 1886, Vol. 5, pp. 408, 411, 412.
39. K. Warren, 'Chemical Foundations, The Alkali Industry in Britain to 1826', Oxford, 1980, p. 47.
40. E. K. Muspratt, op. cit., p. 403, 404, 411.
41. Statute 37 & 38 Victoria c. 43, 'Act to Amend Alkali Act, 1863'.
42. B. P. P. 1878, [c. 2139], XLIV, 1; 1878, [c. 2159-1], XLIV, 43.
43. Statute 44 & 45 Victoria, c. 37, 'Alkali & Works Regulation Act', 1881'.

44. Statute 55 & 56 Victoria c. 30, 'Alkali & Works Regulation Act', 1892'.
45. S.C. Noxious Vapours, para. 1003.
46. Alkali & Works Regulation Act, 1881.
47. B.P.P. 1892, XX, 28th Annual Report of the Inspector of Alkali Works, p. 14.
48. 28th Annual Report of the Inspector of Alkali Works, p. 18.
49. 1st Annual Report of Inspector of Alkali Works, pp. 26-27.
50. 1st Annual Report of Inspector of Alkali Works, p.29 (Estimated from amount of salt used).
51. S.C. Noxious Vapours, para. 1057. 52.
    W.A. Campbell, 'The Old Tyneside Chemical Trade', University of Newcastle-upon-Tyne, 1964, p. 48-50;
    E.E. Aynsley and W.A. Campbell, John Glover and the Clean Air Acts, Chemistry and Industry, 1959, 1540-1541.
53. Hansard, House of Lords, May 9 1862, cols. 1461, 1465.
54. B.P.P. 1840, [c.384] XI, Report from Select Committee on the Health of the Inhabitants of Large Towns, para. 1182.
55. 28th Annual Report of the Inspector of Alkali Works, p. 20.

Death in the Pot: Adulteration of Food and Its Chemical Detection

W. A. Campbell
DEPARTMENT OF CHEMISTRY, THE UNIVERSITY, NEWCASTLE UPON
TYNE NEI 7RU, UK

Food adulteration has a very long history. The Biblical reference
to 'salt that has lost its savour' points to the addition of
tasteless earthy matter in order to cheat the assessors for the
hated tax on salt. Wherever there was a tax on food,drink,spices or
condiments, adulteration might be expected to follow. One kind of
adulteration led to another. For example,milk was often diluted
with water, especially in London where recourse to the 'cow with the
iron tail' was common. Watered milk had a lowered gravity, so this
was corrected by addition of chalk or starch; the flavour was now
poor, so treacle or sugar was added; lastly, some dye was needed to
restore the creamy colour. Thus the original watering led to the
addition of three other substances, and adulteration became more
and more sophisticated.

The movement of people from the country into the towns
associated with the Industrial Revolution brought into being a
system of transport, wholesaling and retailing of food with
consequently enhanced opportunities for adulteration. Not
surprisingly therefore, the second half of the 18th century saw the
publication of a number of books and pamphlets exposing the
nefarious practices of food vendors in the growing towns. Most were
anonymous, and sought to shock rather than to instruct the reader.
Typical of this class was Poison Detected, or Frightful Truths and
Alarming to the Metropolis (1757)[1] .

A very different type of work appeared in 1820. This was
Fredrick Accum's Treatise on Adulteration of Food and Culinary
Poisons exhibiting the Fraudulent Sophistication of bread, beer,
wine, spiritous liquors, tea, coffee, cream, confectionery,vinegar,
mustard, pepper, cheese, olive oil, pickles and other articles
employed in domestic economy, and methods of detecting them. F.C.
Accum came to England in 1793 from his native Germany. He became
engineer to the Chartered Gas Light and Coke Company, lecturer in
chemistry at the Surrey Institution at Blackfriars Bridge, and
librarian at the Royal Institution in Albemarle Street. A prolific
author of textbooks, he had a large private laboratory in London
at which he taught a course in practical chemistry. Returning to
Germany, he was appointed Professors of Chemistry at the Gewerb
Institut in Berlin.[2]

Accum's work differed from its sensational forerunners in two ways; he took his examples not from rumour or hearsay but from law reports; and he gave chemical tests, not always specific, for the adulterants which he described. Moreover, Accum took a personal risk in publishing the names and addresses of brewers, bakers, druggists, tea and coffee dealers, and publicans who had been convicted of fraudulent dealing. In this paper we shall concentrate on three major areas; coloured confectionery; tea, coffee and cocoa; and alcoholic drinks.

1 COLOURED CONFECTIONERY

By 1820 chemical analysis had progressed sufficiently to enable mineral additives to be detected with confidence. The most versatile tool at the analyst's disposal was the blowpipe. Devised by gold- and silver-smiths as a soldering implement, the blowpipe was used by Robert Hooke in 1665 in his examination of petrified wood.[3] J.A. Cramer gave a careful description of the instrument in his Elementa Artis Docimasticae (1739), and references to the blowpipe as a mineralogist's tool are scattered through the work of A.F. Cronstedt (1758). Indeed, Engestrom, Warden of the Mint in Stockholm, stated: "To the best of my knowledge Mr Cronstedt is the first who made such an improvement in its use as to employ it in examining all mineral bodies".

Further improvements and modifications were made to the technique as well as to the instrument by Torbern Bergman, J.G. Gahn, J.J. Berzelius, and J.J. Griffin whose Chemical Recreations (1838) contained the most complete description of blowpipe analysis.[4]

Metallic adulterants were therefore among the first to be systematically investigated. In 1831 The Lancet launched an attack on the sale of sweetmeats coloured with lead, copper, mercury, arsenic and chromium compounds.[5] The tone of the six-page paper was moderate, aiming to acquaint medical men with the results of a careful enquiry into the sale of such goods in London. The editor of The Lancet however was Thomas Wakley, a quarrelsome man with a history of involvement in trouble; indeed, he had founded the journal with the express purpose of combating medical jobbery and nepotism and exposing various kinds of malpractice. He certainly saw the 1831 paper as the first shot in a campaign.

The author, Dr W.B. O'Shaughnessy, being disappointed in his hopes of a London post in forensic medicine, left for India where he eventually became Director of the Electric Telegraph Service. This effectively brought Wakley's campaign to a temporary halt, and not until 1854 was the question of poisonous confectionery again addressed in the pages of The Lancet.

Accum had noticed the presence of copper in sweets. Ostensibly, the green items were coloured with the wholly unobjectionable sap green made from the juice of buckthorn berries, but the shade was often brightened by the introduction of copper. "The foreign conserves such as small green limes, citrons, hop-tops, angelica roots etc., usually sold in round chip-boxes, are frequently impregnated with copper".[6] The same was true of pickles, the presence of copper being betrayed by the red incrustation formed on a steel fork, though in general Accum favoured the ammonia test for copper. His view of the state of food at large was poor: "It would be difficult to mention a single article of food which is not to be met with in an adulterated state, and there are some substances which are scarcely ever to be procured genuine."[7]

The subject of poisonous colours was again explored in 1850 by A. Normandy who ran an analytical, consulting, and teaching practice from his private laboratory in Judd Street, Brunswick Square, London. He reported a case of poisoning at a public dinner in Northampton at which a green blanc-mange had furnished the centre-piece of the table. One man had died and several others had been seriously ill. The coroner's jury found that "Mr Wm Cornfield had died on the 8th of June 1848 from the effects of a poison called arsenite of copper or emerald green, carelessly and negligently administered in a blanc-mange made by Edward Randall, and carelessly and negligently sold by Edmund Franklin with the full knowledge that it contained a deadly poison".[8]

The verdict was less than fair because many people did not in fact know that this pigment, otherwise known as Scheele's Green, was poisonous. Professor A.S. Taylor, Professor of Forensic Medicine at St Mary's Hospital Medical School, reported finding a green stain on a loaf of bread delivered to his own table. The baker had painted his shelves with a bright green arsenical paint, and when Taylor questioned him about the wisdom of this he replied that green paint had never hurt anyone yet.[9] He was by no means alone in his blissful ignorance.

Normandy's tests were made on the dry samples. Arsenic was detected by heating the sample with black flux (fusion mixture plus charcoal) in a glass tube, and observing the mirror of metallic arsenic and noting the odour of garlic. Lead compounds were reduced to a metallic bead by heating on a charcoal block with a blowpipe flame. Chromium gave an emerald green borax bead test. Suspected vermilion was heating in a glass tube with sodium carbonate, and the grey ring of mercury rubbed into silver globules with a glass rod. (Table 1)

The Lancet took up the topic of food adulteration in general in 1851. Publication took the form of reports of the Analytical Sanitary Commission, an entirely unofficial body consisting of Wakley and Arthur Hill Hassall, a physician at the Royal Free Hospital. Previous methods of chemical analysis had largely been restricted to inorganic adulterants, but Hassall applied his great skill as a microscopist, making camera lucida drawings of leaves, dust, insects, starch grains, and crystals. Although of fundamental significance for the examination of food, microscopy was not very useful for detecting poisonous colours; that part of the work was therefore undertaken by Henry Letheby. Letheby was lecturer in chemistry at the London Hospital Medical School, in which post he had succeeded his teacher Jonathan Pereira.

TABLE 1   Poisonous Colours Recognised by Normandy in 1850.

Greens:	Emerald Green) Scheele's Green)	Copper arsenite
	Schweinfurt Green Green verditer Verdigris	Copper Acetate and arsenite Basic copper carbonate Copper acetate
Yellows:	Gamoge Chrome Yellow Orpiment Patent Yellow) Turner's Yellow) Cassel Yellow)	A gum resin Lead Chromate Arsenic sulphide Lead oxychloride
Reds:	Red lead) Minium)	Lead oxide Pb_3O_4
	Cinnabar) Vermilion)	Mercury sulphide
Blues:	Blue verditer	Carbonate of copper and calcium

(Prussian blue was regarded as a safe colour by Normandy)

TABLE 2   Remuneration for Food Analysis in the London Area.

Fulham:	5s to 21s. per sample
Greenwick:	£100 per year
Lewisham:	£50 per year + 10s. per sample
Limehouse:	£150 per year
Plumstead:	£50 per year
Rotherhithe:	2s. 6d. to 10s. per sample
Strand:	£100 per year
Wandsworth:	2s. 6d. to 10s. 6d. per sample
Westminster:	£100 per year
Whitechapel:	10s.6d. to 21s. per sample

Although Letheby continued to use the blowpipe, he added some wet tests. For lead chromate, by far the most widely employed poisonous pigment in confectionery, he boiled the sample with sodium carbonate, filtering off the lead carbonate for blowpipe examination; the yellow filtrate of sodium chromate was acidified to yield the orange colour of dichromate. Scheele's green (copper arsenite) was dissolved in dilute sulphuric acid, and copper detected by means of ammonia solution. For the arsenic he proceeded exactly as Normandy had done four years earlier.[10]

As the case of the green blanc-mange suggests, it was not only in confectionery that the poisonous colours were to be found. Edward Lankester reported a case which occurred in Clifton, Bristol, in 1860 in which Bath buns were coloured with yellow arsenic sulphide to give the impression of a mix that was rich in eggs. London bakers often used turmeric for this purpose, but the Clifton baker had approached an oil and colour-man with a request for a suitable colour. At the enquiry the man said that he was sorry that he had supplied the arsenic colour, but he was temporarily out of lead chromate.[11]

Nor was the danger avoided by choosing only white sweets, for white sugar for confectionery was often mixed with considerable proportions of plaster of Paris. Known in the trade as 'daff' or 'duck', this was frequently confused with white arsenic (III) oxide, sometimes with fatal consequences. Peppermint lozenges seem to have been a common agent of this mode of poisoning.

All the investigators, Accum, O'Shaughnessy, Normandy, Hassall and Letheby were at pains to point out that safer colours were available. But sap green, as Accum had observed, was often adulterated with copper, and annatto was frequently mixed with mercury sulphide (vermilion). By a strange quirk, the vermilion was adulterated with cheaper red lead, which in turn was mixed with the even cheaper red iron oxide, thus making a dangerous situation slightly more tolerable.

The Analytical Sanitary Commission published the names and addresses of the dealers from whom the samples were taken. This had an immediate effect on the London scene from which the worst of the pigments had disappeared in a few years. It had little effect on the rest of the country, where accidents continued to occur. In the closing years of last century an unemployed miner in Houghton-le-Spring, County Durham, began trading in boiled sweets which he sold from a horse-drawn cart in the neighbouring villages. As business improved, he painted the cart (as Princess Charlotte had painted her carriage a century earlier) in lead chromate paint. Pleased with the vivid yellow shade, and having some paint left over, he decided to brighten his sweets in the same way; his good intentions led to widespread illness in the district.[12]

2 TEA, COFFEE AND COCOA

Tea (originally pronounced 'tay' and drunk from dishes or shallow cups without handles) had come into Britain in the middle of the

17th century when the first London coffee-houses were established.
Carried in East India Company ships, it first cost £3/10/0 (£3.50)
per pound; a century later green China tea sold at twelve shillings
(60p) a pound. The tax of five shillings per pound was one of the
major causes of adulteration. The meal of tea and cakes at five
o'clock was introduced by Anna, Duchess of Bedford, because she
said that she had a 'sinking feeling' about that time of day.

Adulterated teas fell into two classes, those adulterated in
the country of origin, and those similarly treated at or near the
point of sale. The teas in the former class were usually
described as 'faced'. The facing consisted of improving the
colour of green teas with Prussian blue, and black teas with
lampblack. Every year two cargoes of Prussian blue, each worth
£2,000, were exported to Canton, to be re-imported into Liverpool
in the form of green tea. The Chinese did not themselves drink
faced tea, but said that since the foreigners preferred it like
that and were prepared to pay high prices, they were willing to
oblige. The quantity of added pigment was usually from one to
three percent, but cases were reported of additions up to ten
percent. The Prussian blue itself was mixed with gypsum which
helped to standardise the shade.[13]

There was much argument as to whether 'facing' amounted to
adulteration. A.H. Allen, Public Analyst for Sheffield and author
of the standard work on commercial organic analysis, held that it
did not. Dr C.M. Tidy was even more cautious: "I should not like
to say very much about it. I am rather disposed to think that it
is one of those things which had better not be interfered with."[14]
It is interesting that Accum had identified verdigris as a facing
agent in green tea.

Although several Acts of Parliament specifically forbade the
practice, other kinds of leaves were often mixed with tea, or even
replaced tea entirely. In 1818 Edward Rhodes was charged with
having "dyed, fabricated, and manufactured divers large quantities,
viz. one hundredweight of sloe leaves, one hundredweight of ash
leaves, and one hundredweight of a certain other tree in imitation
of tea, contrary to the Act of 17 George III." He had in his
possession sloe, ash, elder, and other leaves which were boiled,
baked, rubbed with the hand, and then coloured with logwood. The
colouring matter was easily detected by extracting with water and
treating with sulphuric acid. The problem of identifying foreign
leaves was satisfactorily solved only after Hassall's introduction
of the microscope into food analysis.

More highly publicized were the tea factories in which used
tea leaves were re-constituted. George Phillips of the Inland
Revenue stated in 1843 that there were eight factories in London
alone devoted to this trade.[15]

The used tea leaves were bought from hotels, inns, and
coffeehouses for twopence or threepence a pound. Small amounts of
tannin still remained in the leaves, sufficient to give a dark
colour in the presence of iron. The leaves were therefore dried,
mixed with a little gum, and dusted over with finely ground
copperas (iron II sulphate). The possession of used tea leaves

and iron sulphate in the same establishment was sufficient to merit a prosecution, though it should be remembered that charges were brought for defrauding the Inland Revenue, not for cheating the customer.

In the nature of things, the used tea factory was usually also engaged in passing off foreign leaves. In 1851 premises at Clerkenwell Green were raided and the whole operation laid bare. Several persons were busily employed in making spurious tea. An iron pan suspended before a fire contained sloe leaves with used tea leaves, and in the back room were stored large quantities of used tea leaves, bay leaves, gum, and iron sulphate. Spread on the floor of the drying room were bay and sloe leaves together with used tea leaves, a hundred pounds weight in all. The foreign leaves were easily detected under the microscope, but used tea leaves presented a greater problem; once prepared for sale however, the presence of iron sulphate was a powerful indicator.[16]

Most adulteration took place in wholesale warehouses. The tea came into the ports sealed in lead sheet inside a tea chest, but the excisemen would break a hole in the lead to sample the tea. Through this hole the entire contents of the chest were tipped out spread on the floor, mixed with the chosen adulterant, and poured back into the chest. No workable method for the Customs Officer to seal up the sampling hole was then known.

There were of course scented and flavoured teas which were not adulterated in any real sense, for adulteration implied prejudice to the purchaser. In this category must be placed the tea perfumed with oil of bergamot favoured by that pioneer of Parliamentary reform, Earl Grey.

The position with regard to coffee was a little different in that the Government had permitted the addition of a foreign substance. It was held in some quarters that the addition of a little chicory improved the flavour of coffee. Consequently in 1840 the Chancellor of the Exchequer issued a minute which allowed the dealer to mix chicory with coffee. For some strange reason, no limits were placed on the proportion of coffee which might legitimately be replaced.

The economics of the situation were simple; a grocer was allowed to mix a material valued at eightpence a pound with one worth twice that figure, and to sell the product at the price of the costlier ingredient. Mixtures sold under such names as 'Finest Mocha', 'Choice Jamaica Coffee', 'Finest Java Coffee', 'Delicious Family Coffee' contained never less than one third and often as much as two thirds by weight of chicory. Phillips of the Excise admitted that his men could not determine the quantity of chicory in a sample of coffee, an admission which must have heartened the adulterators. The duty on coffee brought in £45,000 in 1850 so the effect of adulteration in lost revenue must have been serious.

Whatever the shortcomings of the Excise examiners, private firms of analytical and consulting chemists were certainly able

to detect chicory by means of the microscope. This did not deter
Duckworths of Liverpool from taking out a patent for making
imitation coffee beans out of chicory. Eventually the scandal grew
too great for the Government to ignore, and vendors were
compelled to label such coffees as mixtures, though still without
stating the proportions. [18]

The deception did not stop there, for chicory itself was often
adulterated. Among the substances identified by Pereira in
counterfeit chicory were wheat,rye, beans, acorns, carrots,mangold-
wurzel, beetroot, sawdust, burnt sugar (under the trade-name of
Black Jack), and various red earths. A Parliamentary Committee on
food adulteration heard evidence from a Mr Gay in 1855: I remember
one year when chicory was worth £21 per ton, manufacturing 700
tons of carrots into chicory. They were grown by one gentleman in
Surrey, and supplied to the house where I was, and also 350 tons
of parsnips."[19]

Other investigators included spent oak bark from the tanyards.
Roasted and ground peas coloured with red iron oxide were sold to
the trade as 'Hambro powder'.

Chicory apart, there were many other ways of cheapening coffee.
One of the least appealing adulterations involved baked horse
liver. The following description occurs in an anonymous work
entitled Coffee as it is, and as it ought to be: "In various parts
of the Metropolis but more especially in the East, are to be found
liver bakers. These men take the livers of oxen and horses, bake
them and grind them into a powder which they sell to the low-priced
coffeeshop keepers at from fourpence to sixpence per pound, horse-
liver coffee bringing the higher price". This substitution could
be detected by allowing the cup of coffee to stand until cold, when
a thick skin developed on the surface. Both Accum and Normandy
insisted that the only way to obtain good coffee was to purchase a
mill and grind the beans at home.[20]

Cocoa, prepared from the seeds of Theobroma cacao, was
brought from Mexico to Europe by Spanish travellers in 1520; like
tea and coffee, its popularity dates from the late 17th century.
In 19th century England, cocoa was sold under various fanciful
names, 'Soluble Cocoa', Dietetic Cocoa', 'Homoeopathic Cocoa',
all of which were mixtures of cocoa with sugar or starch. The
French Governmnent required such mixtures to be sold as chocolate
and not cocoa.

It is doubtful if the trade regarded these mixtures as being
adulterated, and some claimed that the introduction of sugar or
starch improved the solubility of cocoa. The situation was
further complicated however by the addition of cocoa-nut oil, lard
and sometimes tallow.

Normandy painted a gloomy picture: "Many of the preparations
sold under the names of cocoa flakes, chocolate, or chocolate
power consists of a most disgusting mixture of bad or musty cocoa
nuts, with their shells, coarse sugar of the very lowest quality,
ground with potato starch, oil sea-biscuits, coarse branny flour
and animal fat (generally tallow or even greaves). I have known

cocoa powder made of potato starch moistened with a decoction of
cocoa-nut shells and sweetened with treacle; chocolate made of the
same materials with the addition of tallow and ochre. I have also
met with chocolate in which brick dust and red ochre had been
introduced to the extent of twelve percent; another sample contained
22 percent of iron oxide, the rest being starch, cocoa-nuts with
their shells, and tallows".[21]

Hassall's drawings made the detection of the various kinds of
starch reasonably easy. For animal fats he suggested extraction
with ether and comparison of melting points; here he was making use
of expertise acquired in the study of butter and its adulterations.

3 BEER AND SPIRITS

"There is in this city (London) a certain fraternity of chemical
operators who work underground in holes, caverns, and dark
retirements, to conceal their mysteries from the eyes and
observation of mankind. These subterranean philosophers are daily
employed in the transmutation of liquors, and by the power of
magical drugs and incantations, raising under the streets of
London the coicest products of the hills and valleys of France.
They can squeeze Bordeaux out of the sloe, and draw champagne
from an apple".[22]   This passage of purple prose is Accum's
observation on the subject of alcoholic drinks. It took no account
of the commonest adulteration of these products by the copious
addition of water.

Watered beer loses colour, flavour, and intoxicating power.
Therefore burnt sugar needed to be added for colour, ginger,
coriander or caraway seeds for flavours, and cocculus indicus for
intoxication. The bitterness which ought to have come from hops
was instead supplied by quassia, wormwood or orange peel; cocculus
indicus brought its own bitterness in the form of picrotoxin -
literally 'bitter poison'.

When bitter beer was gaining in popularity in 1850 a report
began to circulate that English brewers added strychnine to
enhance the bitterness. The story stemmed from a remark made by
Payen during a lecture at the Conservatoire des Arts et Metiers.
Commenting on the export of strychnine from Paris to London, he
expressed the opinion that such quantities could only be used in
the brewing industry. As the principal brewers of bitter, Allsopps
at Burton-on-Trent turned to their chemist, H. Bottinger, for
help. Assisted by his former teacher A.W. Hofmann and the Master
of the Mint, Thomas Graham, Bottinger was able to refute Payen's
claim and the whole episode turned into an advertisement for
Allsopp's beer.[23]

Since alcoholic beverages have long been of particular
interest to the Inland Revenue, it is not surprising that they
should have generated some of the earliest laws against
adulteration. An act of Queen Anne forbade the introduction of
cocculus indicus, and colouring with burnt sugar was prohibited
by another of 1817. In the early years of the 19th century
prosecutions were often brought against brewers, publicans, or

brewers' druggists merely for the possession of adulterants on their premises. Dunn and Waller of St John's Street, London, were found in 1817 to possess extract of cocculus indicus, capsicum, iron sulphate, quassia, liquorice, orange powder and ginger, all of which were confiscated.

George Phillips of the Excise testified in 1855 that beers obtained from breweries were consistently stronger than those from publicans, indicating that the watering (with which the whole process of adulteration began) occurred at the point of retail sale. Of twenty samples taken from public houses, fourteen contained grains of paradise to increase the intoxicating power of a watered drink.   With regard to the purity of the malt and hops, the Excise – which employed 70 chemists and 4000 inspectors – showed almost no concern, though Phillips speculated that malt was often cheapened with barley.

The existence of a trade described as 'brewers' druggist' shows that the supply of adulterants was well worth pursuing. As Normandy observed: "Since there are beer druggists, there must be beer druggers".

Powerful Acts of Parliament were certainly on the statute book but they did not succeed in preventing the abuses. This, according to Normandy (who appears to have been prejudiced), was a reflection on the Excise officers. He wrote of "a body of officers, many of whom are celebrated neither for their knowledge or discrimiantion, nor yet for their sobriety, candour or morality." Normandy went on to observe, in words that were to be echoed by the Analytical Sanitary Commission, that the prevention of adulteration ought not primarily to be concerned with protecting the returns due to the Inland Revenue, but instead with safeguarding the health of the consuming public.[24]

The heyday of gin drinking in England occurred around 1746. Hogarth depicted the consequences in his 'Gin Lane' with its caption 'Drunk for penny, dead drunk for twopence'. Gin houses, unlike ale houses, did not have to be licensed with the result that there were some six thousand of them in London. Henry Fielding, from his experience as a magistrate, claimed that gin was the principal sustenance of more than a hundred thousand people in London.[25]

Real 'Holland' gin was spirit distilled from malt barley, rye, and juniper berries. It might be diluted with water which caused turbidity, then fined with a mixture of alum and potassium carbonate known as 'the doctor', which carried down the oily drops. The resulting thin liquor was thickened with sugar, flavoured with cayenne or capsicum, and finally treated with grains of paradise.

This procedure applied when the starting material was genuine gin. There were however several manuals in circulation which described the production of spurious gin. A 'Plymouth' gin might be made as follows:

```
700 gallons of spirit
 14 pounds of German juniper berries
  1½ pounds of calamus root
  8 pounds of sulphuric acid
```

British gin was made in a similar manner:

```
120 gallons of rectified spirit
  ½ ounce of oil of vitriol
  ½ ounce of oil of almonds
  ½ ounce of oil of turpentine
  1 ounce of juniper berries
  1 gallon of lime water
  1 gallon of rose water
 25 pounds of sugar dissolved in
  9 gallons of Thames water
```

The presence of sulphuric acid was detected by evaporating the spirit to a small volume, when charring of the organic matter occurred.[26]   The publication in The Lancet of the reports of the investigations of Hassall and Letheby caused such a stir that the Government felt compelled to set up its own Parliamentary Committee on Adulteration in 1855. The Committee stated in its report :
"Not only is the public health thus exposed to danger, and pecuniary fraud committed on the whole community, but the public morality is tainted, and the high commercial character of the country seriously lowered both at home and in the eyes of foreign countries."

In 1860 there was passed an Act for Preventing the Adulteration of Articles of Food and Drink, which permitted but did not compel a borough to appoint a Public Analyst. This Act was so defective that it failed almost totally to fulfil its intentions, though one of the very few Public Analysts appointed under its terms was Henry Letheby.[27]

A second Act of 1872 required those boroughs which were police authorities to appoint analysts, and stipulated that the analyst should be competent in chemical, medical, and microscopical investigation. The only system of chemical qualification which existed at that time was that administered by the Pharmaceutical Society (founded in 1841), and several prominent pharmacists did become Public Analysts.[28]

If the Public Analyst's findings were challenged by the vendor an independent analysis was to be sought. But if the two analysts did not agree, the final arbiter was to be 'Somerset House' – in other words the chemists of the Inland Revenue. These were men who were skilled in determining alcohol in drink or moisture in tobacco, but who possessed no wider experience in chemical analysis; to make matters worse, Somerset House refused to divulge the analytical procedures used there.[29]   Partly as a result of these galling conditions, the Public Analysts came together in 1874 to form their own society; after many metamorphoses this body was to become the basis of the Analytical Division of the Royal Society of Chemistry.

The analysts were in general poorly paid (Table 2) thus increasing
the tendency for the posts to be held on a part-time basis by
Medical Officers of Health or by University professors of chemistry.
Moreover it was alleged that the Local Boards who made the
appointments were largely composed of tradesmen who would not
appoint a visibly enthusiastic and competent candidate. Certainly
some of the questions put to applicants seem to justify this
conclusion.[30]

In order to provide a more relevant guide to professional
competence than either a degree in medicine or Membership of the
Pharmaceutical Society, the Institute of Chemistry was founded in
1877. Its early years were plagued by controversy, and its
relations with the Public Analysts were not at first wholly cordial.
By the slow exercise of diplomacy, and a modicum of cunning, the
many problems were overcome but it was not until 1900 that the
Local Government Board issued a suggestion that the conditions of
the Foods and Drugs Act would be met if the applicant possessed
either the Fellowship or the Associateship of the Institute of
Chemistry of Great Britain and Ireland.[31]

This explains in part how the situation described in the
Quarterly Review could persist for so long. "Like a set of monkeys
every man's hand is seen in his neighbour's dish. The baker takes
in the grocer, the grocer defrauds the publican, the publican
'does' the pickle-maker, and the pickle-maker fleeces and poisons
all the rest."[32]

REFERENCES

1.   C.A. Mitchell, "Forensic Chemistry in the Criminal Courts",
      Institute of Chemistry Lecture Reprint 1938, 11,12.
2.   J.R. Partington, "History of Chemistry", Macmillan,London
      1962, vol. 3. 827.
3.   R. Hook, "Micrographia", Martyn & Allestry, London 1665,108.
4.   W.A. Campbell, Proc.Univ. Newcastle Phil.Soc., 1971-72, Vol.2,
      No. 2: J.J. Griffin, "Chemical Recreations", Griffin,Glasgow,
      8th ed. 1838, 167.
5.   W.B. O'Shaughnessy, The Lancet, 1830-31 (ii), 193.
6.   F. Accum, "TReatise on Adulteration of Food", Longman,London
      1820, 317.
7.   Abid., 3.
8.   A. Normandy, "Commercial Handbook of Chemical Analysis",
      Knight, London 1850, 78.
9.   A.S. Taylor, Medical Times & Gazette, 1854,n.s.8 326.
10.   The Lancet, 1854 (i), 316, 428, 524,581, 1858 (ii),536,636,661.
11.   W. Crookes, Chemical News 1860, 1, 48.
12.   E. Gabriel Jones, "Foods Fakds: Ancient and Modern",Institute
      of Chemistry Lecture Reprint, 1930, 16.
13.   J.F.W. Johnston "The Chemistry of Common Life" (rev.A.H. Church)
      Blackwood, London 1891, 134.
14.   A.H. Hassall, "Food, its Adulterations and the Methods for
      their Detection", Longman, London 1876, 258
15.   Ibid., 136
16.   "   138
17.   "   158

18. Ibid., 156; A.H. Allen, Analyst, 1880, 5, 1, 87, 227.
19. Ref. 14, 183.
20. Ref. 6, 290.
21. Ref. 8, 179; J. Muter, Analyst, 1879, 4, 65.
22. Ref. 6, 98.
23. C.A. Russell, N.G. Coley and G.K. Roberts, "Chemists by Profession", Open Univ. Press, Milton Keynes 1977, 37.
24. Ref. 8, 62.
25. G.M. Trevelyan, "Illustrated English Social History", Penguin, Harmondsworth 1964, 88-90; D. George, "England in Transition", Penguin, Harmondsworth 1962, 68.
26. Ref. 14, 810-812.
27. 23 & 24 Victoria cap. 84.
28. 35 & 36 Victoria cap. 74
29. B. Dyer and C.A. Mitchell, "Fifty Years of the Society of Public Analysts", Heffer, Cambridge 1932, 2; R.C. Chirnside and J.H. Hamence, "The Practising Chemists", Society for Analytical Chemistry, London 1974, 68.
30. C. Alder Wright, Medical Circular, 1876 (ii), 374; J. Soc. Arts, 1874, 23, 434.
31. Report of Local Government Board, 1900-01, 30, app. 2.
32. Quarterly Review 1855, 96, 493.

Succès et Problèmes Rencontrés par les Produits Agrochimiques

D. Demozay

ENVIRONMENTAL CHEMISTRY, RHÔNE-POULENC AGRO, 14/20 RUE
PIERRE BAIZET, 69009 LYON, FRANCE

I   INTRODUCTION

Les produits agrochimiques sont avant tout destinés à prévenir les dégats causés aux cultures par les animaux nuisibles, les maladies et les mauvaises herbes. Ils contribuent ainsi à la mise à disposition permanente des populations d'une nourriture abondante, variée et saine.

Durant les quarante dernières années, la population mondiale est passée de 2, 5 à plus de 5 milliards d'habitants. Sans l'utilisation de méthodes modernes de production, et notamment sans la mise au point et l'emploi de produits agrochimiques bien adaptés, il eut été impossible de faire face à ce doublement. Rien ne permet de prévoir que cette situation puisse changer rapidement. On prévoit que la population mondiale passera le cap des 8 milliards aux environs de l'an 2020.

2   LES SUCCES

Parmi les succès obtenus au fil des années, nous en retiendrons cinq que nous illustrerons d'exemples pris dans les domaines insecticides, acaricides et fongicides, étant bien entendu que les herbicides auraient permi de tirer les mêmes conclusions.

La découverte de matières actives dans des familles nouvelles

Les insecticides les plus anciens étaient d'origine végétale (nicotine, pyrèthre, roténone, quassine...) ou d'origine minérale (arsenicaux, fluorés, polysulfures de chaux...).
Une première série d'insecticides organiques de synthèse vit le jour avec la découverte des propriétés insecticides du DDT par le suisse Paul Muller en 1939. Mis sur le marché en 1942, cet insecticide fit ses premières preuves en enrayant une épidémie de typhus à Naples en 1943. L'aldrine, l'heptachlore, le toxaphène qui datent tous de 1948 font partie de cette première génération d'insecticides tombés en désuétude à partir des années soixante dix et pratiquement retirés du marché dans tous les pays tempérés.
Le DDT fut interdit en France entre 1972 et 1974.
Le lindane (1942) et l'endosulfan (1956) restent par contre utilisés dans des créneaux spécifiques ou leurs propriétés permettent d'obtenir d'excellents résultats.

Les insecticides organophosphorés découverts pour les premiers en 1940, ont des propriétés très diverses, ce qui explique l'extraordinaire succès de cette vaste famille. On peut en rapprocher les carbamates, qui agissent comme les organophosphorés en inhibant les cholinestérases du système nerveux des insectes. Le tableau 1 rappelle les principaux insecticides dont les marchés mondiaux en 1989/1990 dépassaient 50 millions de dollars.

Depuis des siècles, on utilisait des extraits de fleurs de *Chrysanthemum pyrethrum* dans la lutte contre les insectes. Il était impossible de les utiliser économiquement dans le domaine agricole, en raison de leur destruction très rapide par la lumière. L'allethrine synthétisée en 1949 ou la phénothrine (1973) permettaient de s'affranchir de la culture du pyrèthre, sans améliorer toutefois la photolabilité. En 1973, au moment ou disparaissait le DDT, Elliott réalisait la synthèse de nouveaux dérivés cent fois plus actifs que le DDT et cette fois stables à la lumière ce qui leur ouvrait l'accès du monde agricole. On peut ainsi distinguer quatre étapes dans la mise au point des pyréthroïdes :

1) les premiers insecticides photolabiles (alléthrine, tétraméthrine)
2) les produits photostables (fenvalérate, perméthrine, cyperméthrine)
3) des produits plus élaborés, enrichis en isomères ou portant des substitutions sur les moitiées alcool ou acides (lambda-cyalothrine, alpha-cyperméthrine...)
4) des produits plus récents doués d'un spectre d'activité original (tefluthrine, bifenthrine)

La chute des brevets des pyréthroïdes de seconde génération, et l'apparition de résistances dans certains groupes d'insectes a conduit à rechercher, et à trouver des insecticides actifs dans de nouvelles familles chimiques. C'est le cas des pyrroles expérimentales d'American Cyanamid (AC 303, 630) et d'une nitroguanidine, l'imidaclopride de Bayer.

Dans ce contexte, Rhône-Poulenc a récemment annoncé son intention de développer une nouvelle famille de produits insecticides et acaricides actifs à très faible dose, les "fiproles". Le premier représentant de cette famille , le fipronil, a fait l'objet d'une demande d'autorisation de vente en 1992.

Tableau 1 Principaux insecticides sur le marché mondial (1989/1990)

Famille	Nom	année	marché (10^6 \$)
OC	endosulfan	1956	110
OP	chlorpyriphos-et.	1965	305
OP	monocrotophos		110
OP	parathion	1946	82
OP	méthamidophos	1970	80
OP	chlorpyriphos-met.	1966	70
OP	profenophos	1975	65
OP	azinphos-me.	1955	61
OP*	disulfoton		60
OP*	fonofos	1965	59
CA*	aldicarb	1965	181
CA	méthomyl	1968	112
CA	carbaryl	1957	86
CA	thiodicarb	1977	56
CA*	carbofuran	1965	56
PY	deltaméthrine	1974	235
PY	cyperméthrine	1975	85
PY	perméthrine	1973	64
PY	cyalothrine	1980	63
PY	fenvalérate	1974	63
PY	cyfluthrine	1981	60

* traitement du sol

Les acaricides appartiennent soit à des familles chimiques connues par ailleurs pour leurs propriétés insecticides, soit à des familles originales regroupées sous le terme d'acaricides spécifiques. Delorme et Dacol (1988) ont résumé les principaux modes d'action des acaricides et les phénomènes de résistance associés. On sait que les acariens, comme certains insectes, sont susceptibles de développer des résistances aux produits utilisés. Ceci est particulièrement net lorsqu'une forte pression de selection est exercée sur des ravageurs ayant un cycle de reproduction rapide. Les résistances constatées dans le groupe des acariens expliquent en grande partie l'érosion rapide des produits et les nombreux retraits d'acaricides spécifiques constatés durant ces dernières années (chlorbenside, chlordimeform, chlorobenzilate, chloropropylate, cyhexatin, fénazaflor, thioquinox, etc.).

La multiplicité des mécanismes de résistance et la complexité des modifications développées par les acariens à la suite des traitements laisse penser que toutes les espèces sont à terme capables de développer une résistance à quelque acaricide que ce soit, à moins de prendre justement en compte ce phénomène dans les stratégies d'utilisation. La diversité des familles d'acaricides constitue actuellement l'arme la plus efficace pour lutter contre les souches résistantes. En 1987, sur 45 acaricides autorisés en France, 23 également autorisés comme insecticides appartenaient aux grandes familles d'insecticides classique : organophosphorés, carbamates, pyrethroïdes, mais les 22 autres appartenaient à d'autres familles.

Au cours des années 80, les nouveaux acaricides apparus dans le commerce appartenaient souvent à des familles entièrement originales :

Pyrethrinoïdes	fenpropathrine	(1982)
Tetrazines	clofentézine	(1984)
Thiozalodines	flubenzimine	(1984)
Thiozalodinones	hexythiazox	(1985)
Avermectines	abamectine	(1985)
Norpyrethrates	acrinathrine	(1989)
Acylurées	flufénoxuron	(1989)

Les acaricides en cours de développement au début des années 90 confirment cette tendance à rechercher une activité dans des familles chimiques originales, qui compensent leur coût de fabrication élevé par des doses d'utilisation plus faibles :

			g/hl
Benzoylurée	flufenoxuron	Shell	5-10
Benzolylurée	flucycloxuron	Duphar BV	12. 5
	acrinathrine	Roussel-Uclaf	6-7. 5
	pyridaben	Nissan	15
Phenoxypyrazole	fenproxymate	Nihon-Nohyaku	2. 5-5
Pyrazole	MK 239	Mitsubishi	5-20

La découverte de produits efficaces à des doses de plus en plus faibles

Indépendament de la diversification des structures chimiques, on constate que les molécules récentes sont souvent beaucoup plus efficaces que les produits anciens et donc actives à des doses réduites. Delorme (1989) a bien montré cette tendance pour les insecticides homologués en France contre les pucerons , les acariens et contre le carpocapse. Dans le cas des traitements du sol, l'emploi de nouveaux insecticides, combiné avec la localisation des traitements sur la ligne de semis, a permi de diminuer les doses de 2,5 -12,5 kg/ha à 0,1 kg/ha.

Les céréales, les arbres fruitiers et les cultures légumières demandent une forte protection contre les maladies cryptogamiques. La lutte contre les maladies fongiques des plantes remonte à 1807, lorsque Benedict Prévost recommanda l'utilisation du sulfate de cuivre pour empècher la contamination des semences de blé par la carie.

Les champignons, qui se développent sur les végétaux d'autant plus rapidement que le climat est chaud et humide, vivent en parasites de leurs hôtes et il est donc très difficile de les maîtriser une fois installés. Pour cette raison, les premiers fongicides découverts devaient s'appliquer de façon préventive. C'était par exemple le cas des produits à base de cuivre, utilisés pour la protection des vignes contre le mildiou depuis 1880 (Bouillie Bordelaise). Les fongicides de synthèse moderne ont été découverts dans de très nombreuses familles chimique. Les plus grands progrès ont été obtenus avec 1) la mise au point de produits curatifs,et 2) la découverte de fongicides préventifs systémiques à longue durée d'action.

De très nombreux fongicides efficaces contre les maladies des céréales appartiennent à la famille des triazoles. Le dernier commercialisé de cette série, le bromuconazole, autorisé à la vente en France dans le courant de l'année 1992, agit en inhibant la biosynthèse de l'ergostérol au niveau de la C14 déméthylase, ce qui provoque chez les champignons une perméabilité des membranes cellulaires. Fongicide à spectre large, il protège les céréales contre l'ensemble des maladies des tiges et du feuillage, comme l'oïdium, les rouilles, la septoriose, la rhynchosporiose et même le piétin-verse. Il est en plus applicable sur d'autres cultures comme les arbres fruitiers, les cultures légumières et les bananiers qu'il protège contre la cercosporiose.

Les innovations en agrochimie résultent des progrès réalisés dans la compréhension des voies de la biosynthèse spécifique des plantes, dans celle des mécanismes de pénétration et de transport dans les plantes et dans une meilleure connaissance de la biologie et de la physiologie des semences. C'est ainsi que très récemment Rhône-Poulenc a annoncé la mise au point du triticonazole, un fongicide qui appliqué en traitement de semences assure une longue protection contre les maladies des tiges (fusariose et piétin) et contre celles du feuillage, traditionellement combattues par des pulvérisations foliaires en cours de végétation. Ce type de protection par la semence 1) facilite l'implantation de la culture, 2) minimise l'impact du produit en dehors de la rhizosphère, 3) réduit les manipulations et les passages en champ.

Cette tendance à la diminution des doses à l'unité de surface traitée est générale, il serait facile de la démontrer également à l'aide d'exemples pris parmi les herbicides.

La prise en compte très précoce de l'environnement et de la toxicité dans les tests de selection.

Avant qu'un produit agrochimique n'atteigne le stade commercial, il doit franchir un nombre considérable de tests destinés à prouver son efficacité ainsi que l'absence pratique de risque pour l'homme et l'environnement dans les conditions d'emploi qui seront recommandées. Ces tests, définis par des textes réglementaires, sont maintenant réalisés conformément aux Bonnes Pratiques de Laboratoire (BPL). Il faut compter de 7 à 10 ans pour la réalisation de l'ensemble des essais qui seront rassemblés dans le dossier d'homologation remis aux services officiels.
La Directive du Conseil du 15 juillet 1991, concernant la mise sur le marché des produits phytopharmaceutiques, parue sous le n° 91/414/CEE au J. O. des Communautés européennes du 18. 8. 91, dresse dans ses annexes II et III la liste des études exigées avant de pouvoir commercialiser un produit dans les pays de la Communauté. Cette liste résulte de l'expérience acquise durant une cinquantaine d'années et constitue la somme des connaissances scientifiques disponibles à ce jour pour apprécier les diverses propriétés d'un produit phytopharmaceutique.

L'importance prise par les exigences réglementaires a obligé les industriels à concevoir des batteries de tests extrèmement précoces pour éliminer le plus rapidement possible les molécules efficaces, mais qui ne répondent pas aux critères exigés pour la protection des utilisateurs, des consommateurs et de l'environnement.

Les substances qui ont été homologuées ces dernières années, ainsi que les substances qui le seront dans les temps à venir auront donc dès le départ été conçues pour respecter l'environnement. On se souviendra que la législation européenne, comme la législation américaine, prévoit la réhomologation de tous les produits au bout d'une période de 10 ans. Il en résulte que les dossiers d'homologation sont constament remis à jour et que seuls peuvent subsister les produits répondant aux critères de sécurité les plus récents.

La mise au point de formulations mieux adaptées.

Il est impossible d'utiliser les matières actives en l'état. Il faut :

- diviser la matière active pour assurer une bonne couverture de la cible,
- optimiser les performances biologiques de la (des) matière(s) active(s),
- faciliter la mise en oeuvre par l'agriculteur,
- assurer la stabilité chimique de la matière active,
- assurer la sécurité de l'emploi.

C'est le rôle de la formulation que de présenter la substance sous une présentation qui améliore l'effet recherché (dose efficace, sélectivité, largeur du spectre...) et de diminuer les inconvénients possibles (phytotoxicité, persistance, mobilité dans le sol, toxicité, coût...). En se limitant au seul cas des formulations appliquées sous forme de bouillie, on constate qu'en partant de concentrés émulsionnables (CE), utilisant de grandes quantités de solvants, ou des poudres mouillables (PM), génératrices de poussières, l'agrochimie a su développer de nombreuses présentations qui n'ont pas ces inconvénients : sachets hydrosolubles, micro-émulsions (ME), émulsions concentrées (EC), suspo-émulsions (SE), granulés dispersables (GD), suspensions concentrées (SC), etc.

Une très grande sécurité pour le consommateur.

Beaucoup de produits phytosanitaires ne laissent pas de résidus décelables au moment de la récolte. Néamoins, des Limites Maximales de Résidus (LMR) sont fixées par le législateur pour toutes les substances autorisées et ceci pour toutes les denrées consommées. Pour vérifier que les agriculteurs et que les importateurs d'aliments respectent ces limites, de nombreux pays établissent des programmes de surveillance pour détecter et mesurer les résidus dans les fruits, les légumes, les céréales, etc.

Le Codex Alimentarius détermine pour toutes les denrées circulant entre états et pour toutes les substances susceptibles d'être présentes au moment de la récolte, des LMR valables dans le cadre des échanges internationaux. Les USA et la CEE fixent également des LMR qui doivent être respectées par les états relevant de leur souveraineté.

Une LMR tient compte des données toxicologiques, mais ne représente pas une limite toxicologique à ne pas dépasser. Elle constitue une norme permettant de s'assurer que le produit a bien été utilisé dans les conditions prescrites par les autorités.

La quantité d'un produit agrochimique à laquelle nous sommes susceptibles d'être exposés sans risque en consommant notre ration alimentaire quotidienne est calculée à partir des nombreuses données expérimentales existantes. L'Organisation Mondiale de la Santé (OMS) fixe ainsi une Dose Journalière Acceptable (DJA) qui peut servir de référence.

Pour simplifier, nous admettrons que la DJA indique les quantités de résidus d'un produit donné, provenant de toutes les sources possibles, que nous pouvons consommer tous les jours de notre vie sans ressentir d'effet néfaste.

La Food and Drug Administration (FDA) chargée de la surveillance des aliments aux USA a publié les résultats des dosages effectués en 1990 (J. Assoc. Off. Anal. Chem., 74, September-October, 1991). Ces résultats regroupés dans le tableau 2 montrent qu'une immense majorité de denrées ne contient pas de résidus détectables ou ne contient que des résidus conformes aux MRL et qu'une proportion très faible contient des résidus non conformes, soit parce qu'une LMR est dépassée, soit parce que le produit retrouvé ne dispose pas de LMR aux USA (ceci est vrai principalement dans le cas des denrées importées).

Tableau 2   Résultat des analyses effectuées en 1990 par l'USDA (% des échantillons dans chaque catégorie)

	0 résidus	Résidus < LMR	Non conforme
Denrées domestiques			
Grains et dérivés	54,4	44,8	0,8
Lait et dérivés/oeufs	90,7	9,3	0
Poissons/coquillages	31,5	67,4	1,1
Fruits	51,2	48,4	0,4
Légumes	62,5	35,7	1,8
Divers	78,5	21,2	0,3
Denrées importées			
Grains et dérivés	55,7	34,5	9,8
Lait et dérivés/oeufs	92,1	7,9	0
Poissons/coquillages	89,1	10,9	0
Fruits	62,8	35,1	2,1
Légumes	61,5	33,8	4,7
Divers	70	21,1	8,9

A partir des résultats des dosages effectués en 1989 (J. Assoc. Off. Anal. Chem., 73, September-October 1990) l'USDA a calculé pour près de 190 produits les quantitées ingérées chaque jour avec leurs aliments par trois sous groupes de la population, regroupés en fonction de l'âge et du sexe. Ces quantités ont été comparées aux DJA correspondantes avec les résultats suivants :
    I50 produits n'ont pas été détectés
    21 ont été détectés en quantité < 0,1% de la DJA
    16 ont été détectés en quantité comprise entre 0,1 et 1%   de la DJA
    4 ont été détectés en quantité comprises entre 1 et 5% de la DJA

Toutes les études conduites en Europe durant ces dernières années aboutissent à des résultats très voisins de ceux des USA. Ils démontrent que l'agrochimie a su mettre au point des produits et des techniques d'application d'une très grande sécurité pour les consommateurs. Ils apparaîtraient encore bien plus satisfaisants si l'on savait les mettre en parallèle avec les énormes quantités de toxines naturelles aux propriétés toxicologiques quasi inconnues présentes dans nos aliments. Ames et Gold (1991) estiment que " près de 99,99% des pesticides présents dans l'alimentation humaine sont des pesticides naturels produits par les plantes. Toutes les plantes produisent des toxines pour assurer leur protection contre les champignons, les insectes et les prédateurs animaux tels que l'homme. Des dizaines de milliers de ces pesticides naturels ont été découverts, et chaque espèce de plante contient sa propre série de toxines diverses, généralement au nombre de quelques douzaines. Lorsque les plantes sont stressées, ou endomagées (par exemple à l'occasion de l'attaque d'un ravageur) elles augmentent plusieurs fois leur teneur en pesticides naturels, éventuellemnt jusqu'à des niveaux toxiques pour l'homme. Nous estimons que les américains ingèrent près de 1500 mg de pesticides naturels par personne et par jour, ce qui représente 10 000 fois ce qu'ils consomment sous forme de résidus de pesticides de synthèse...."

3   LES PROBLEMES

Dès son origine, l'agrochimie a dû faire face à des problèmes tels que la variation saisonnière des utilisations liée à la croissance des cultures, ou tels que la difficulté des prévisions dans un marché totalement soumis aux conditions climatiques. En cela elle partage les soucis des limonadiers et des marchands de parapluies. Mais l'Agrochimie rencontre aujourd'hui des problèmes spécifiques, souvent plus politiques, réglementaires, éthiques ou de communication que purement scientifiques, qui tous présentent de sérieuses répercussions économiques.

La 9ème édition du "Pesticide Manual" (BCPC, 1991) nous renseigne sur plus de 600 substances encore commercialisées dans le monde, mais entre 1987 et 1991, le nombre des substances abandonnées car dépassées techniquement s'est élevé de 327 à 486.

Parmi les challenges auxquels l'Agrochimie doit faire face, nous prendrons sept exemples :

Les conséquences de la redéfinition de la politique agricole communautaire (PAC) en Europe.

L'extrême disparité des situations existant entre les pays en ce qui concerne les surfaces disponibles par habitant et les pourcentages de ces surface affectés aux cultures sont à l'origine de coûts de production extrèmement divers. L'Europe, avec des surfaces cultivées limitées, a soutenu les prix des denrées afin de conserver la maîtrise de ses sources de nourriture obtenue de façon intensive. D'autres pays comme les USA, l'Australie ou le Canada ne sont pas soumis aux mêmes contraintes et ont de ce fait des prix de production beaucoup plus faibles. La CEE a décidé de rapprocher les prix européens de ceux du marché international. Il s'agit d'une décision politique qui se traduira par une forte baisse de revenu des agriculteurs. Par exemple, le prix indicatif des céréales est fixé à 130, 120, et 110 écus/t pour les campagnes de 1993/94, 1994/95 et 1995/96, respectivement. Il est évident que de telles mesures auront des répercussions sur le marché des produits agrochimiques.

Il paraît toutefois que les zones les mieux adaptées aux cultures continueront à pratiquer une agriculture intensive faisant appel largement à l'agrochimie. Les produits utilisés seront plus que jamais choisis en fonction de leur capacité à respecter l'environnement. Agriculture intensive et respect de l'environnement ne sont nullement incompatibles, mais nécessiteront de gros efforts d'innovation et de développement.

La mise en place de réglements dépourvus de justifications scientifiques.

La mise en place de réglementations pour tenir compte d'une estimation des risques et des avantages présentés par une solution à un problème donné est justifiée dans la mesure ou elle est basée sur une connaissance objective des faits. Dans le domaine de l'agrochimie, ce n'est pas toujours le cas. Certaines normes fixées de façon arbitraire représentent des contraintes soit impossibles à respecter, soit économiquement si lourdes qu'elles risquent de drainer tout le potentiel de Recherche vers la résolution de problèmes inexistants.

Par exemple la Directive CEE relative à la qualité des eaux destinées à la boisson a fixé pour tous les pesticides une même norme de pureté de 0, 1 µg/l, sans tenir compte de la diversité des toxicités des substances. Cette valeur a été retenue parce que les experts la considérait à l'époque comme la limite de détection des résidus.

1)   Cette mesure est injuste à l'égard des produits agrochimiques. Rien ne permet de justifier le regroupement des pesticides dans une même catégorie alors que les autres produits chimiques sont classés séparément et que par exemple le cyanure (CN) reste toléré à la concentration de 10 µg/l (Arrêté français du 20 février 1990).

2)   Cette mesure est absurde. L'OMS propose des limites calculées à partir de la DJA : une personne pesant 70 kg pourrait absorber 10% de la DJA dans deux litres d'eau tous les jours de sa vie. L'Environmental Protection Agency

(EPA) propose, aux USA, un calcul similaire avec 20% de la DJA. Le tableau 3 présente quelques valeurs publiées par l'US-EPA en Janvier 1989. On voit que pour les produits actuellement autorisés en France, les tolérances aux USA sont de 20 à 21 000 fois supérieures à la limite européenne. Si l'on se souvient que la DJA est elle même calculée avec une marge de sécurité de 100 par rapport aux doses sans effet chez les animaux de laboratoire et que l'EPA ne prend en compte que 20% de la DJA dans son calcul, on constate que les marges exigées par la CEE par rapport aux doses sans effet chez les animaux vont de 10. 000 à plus de 10 millions. (Si l'on admet que la "dose sans effet" entre deux voitures est de 30 mêtres, une marge de 10. 000 reviendrait à exiger sur les autoroutes le respect d'un intervalle de 300 kilomêtres entre deux véhicules).

Tableau 3 Concentrations maximum tolérées dans l'eau aux USA exprimée en µg/l (Office of Water, U. S. EPA, January, 1989)

Concentration	Substance
2	lindane, methyl-parathion
3	atrazine
3,6	MCPA
10	aldicarb, diuron
40	carbofuran
70	2,4-D
200	methomyl, metribuzine
700	carbaryl, glyphosate
2100	dimethrine

3)   Imaginons qu'un produit soit autorisé sur cerisier et que le législateur ait fixé la LMR à 1 ppm (1 mg/kg). En lavant une cerise conforme à la norme, donc considérée comme parfaitement saine et comestible, on obtiendrait environ 30 litres d'eau ne respectant pas la norme des eaux destinées à l'alimentation humaine (soit la quantité suffisante pour un adulte pendant 15 jours).

4)   Certains pays européens voudraient étendre la norme à l'ensemble des eaux de surface, or il suffirait de 20 mg pour élever à 0, 1 µg/l la concentration de toute l'eau tombée sur un hectare lors d'une pluie de 20 mm. On conçoit bien que dans les régions agricoles, cette norme n'est pas réaliste quelles que soient les mesures prises.

5)   La norme européenne, dont on voit le caractère arbitraire, coûte extrêmement cher en imposant le traitement préalable des eaux destinées à la boisson et en mobilisant les ressources de l'industrie et des collectivité dans des programmes de mise au point des méthodes d'analyse et de surveillance du respect des normes.

L'augmentation du coût du développement d'un produit.

Les études réglementaires nécessaires pour pouvoir développer une nouvelle substance ont vu leur coût augmenter considérablement au cours des dernières années. Ceci est dû à la fois au plus grand nombre d'études exigées et à la mise en place des bonnes pratiques de laboratoire (BPL).

Le GIFAP (1990) au moment où la Directive 91/414/CEE était encore à l'état de projet, avait évalué le coût de réalisation des études pour l'homologation européenne d'une nouvelle matière active, appliquée sur une seule culture, à plus de 70 millions de Francs français (7, 1 millions de £ sterling). Si l'on ajoute à ces frais, le coût de la recherche, de la découverte et de l'évaluation des molécules avant la décision de développement, ainsi que celui des études de procédés et l'investissement industriel, on comprend que seuls quelques rares objectifs justifient un tel effort économique.

Il en résulte également que seules quelques très grandes sociétés internationales peuvent encore espérer découvrir et développer de nouvelles molécules.

L'augmentation de la durée des études avant commercialisation.

La réalisation séquentielle des études de toxicité aiguë, des études subchroniques, puis des études chroniques, de même que la mise en place d'essais de persistance en plein air, nécessite plusieurs années. A cela s'ajoutent les délais d'examen des dossiers, ainsi que le temps de réponse aux questions possibles des examinateurs. On admet qu'il s'écoule de 7 à 10 ans entre le moment ou une molécule nouvelle est synthétisée, et le moment ou elle arrive sur le marché. Compte tenu des frais engagés pendant toute cette période, on ne peut espérer atteindre l'équilibre entre les recettes et les dépenses avant la treizième ou la quatorzième année. La période de protection des brevets étant en moyenne de 20 ans, on voit que l'Agrochimie, comme la Pharmacie, ne dispose que d'un très faible nombre d'années pour valoriser ses découvertes.

Il en résulte que la revalorisation de la durée de protection des brevets, constitue un élément majeur de la survie de l'industrie agrochimique.

Le manque de modèles de prévision du comportement des produits

Le risque présenté par un produit phytosanitaire à l'égard de l'environnement dépend de ses transformations (métabolisme), de la durée d'exposition et des concentrations dans l'écosystème. Il faudrait prévoir avec précision ce que deviendra une substance appliquée dans certaines conditions.

Le coût des études sur le terrain, réalisées à l'échelle d'un bassin versant ou d'un secteur plus limité (mésocosme), devient vite prohibitif, dès que l'on s'écarte de la dimension du champ. Un grand nombre d'approches expérimentales sont donc proposées pour évaluer le comportement des produits dans l'environnement d'une façon plus accessible en faisant appel soit au laboratoire, soit à des essais à petite échelle (ex. : lysimètres), soit à des modèles mathématiques qui prennent en compte les propriétés physicochimiques caractéristiques de chaque matière active : solubilités, tension de vapeur, stabilité à l'hydrolyse et à la photolyse, coefficients de partition eau/octanol, KH, KOC, KOM, etc. Ces modèles informatisés permettent théoriquement de réaliser des simulations et donc d'établir la probabilité des contaminations dans des circonstances données.

Une étude du GIFAP (1992) montre les limites actuelles de deux approches : d'une part celle qui fait appel à des lysimètres, d'autre part celle qui utilise la modélisation du déplacement des produits dans le sol :

Lysimètres
La taille des lysimètres limite le choix des cultures.
Les lysimètres présentent une zone saturée qui n'existe
généralement pas dans les conditions réelles.
Les lysimètres surestiment les concentrations dans le percolat.
Des imperfections structurelles (fissures du sol) peuvent entrainer des
surestimations considérables.

Modèles
Il est très difficile de vérifier la validité des modèles.
Beaucoup de modèles sont utilisés sans validation adéquate.
L'utilisation de données expérimentales médiocres invalide les résultats.
Les hypothèses faites lors de la conception du modèle ont une influence
sur les résultats.
La selection des paramètres modifie les résultats.
Les concentrations calculées ne correspondent pas à la réalité.

Actuellement des études se poursuivent en vue de mettre au point, d'améliorer ou de valider un grand nombre de modèles :

- Modèles de type universel (McKay) permettant de prévoir quantitativement dans quels compartiments un produit se trouvera après une longue période d'utilisation (à l'équilibre).
- Modèles de type local (PRZM) qui permet à partir de paramètres hydrologiques, pédologiques et climatiques d'un terrain, de savoir si un produit donné atteindra la nappe phréatique.
- Modèles hydrologiques (DISPERSO de la Compagnie Générale des Eaux) permettant pour un cours d'eau étalonné, de prévoir le comportement du produit tout au long du flux lors d'un apport ponctuel.
- Modèles de Bassin Versants, tels que ceux étudiés dans le projet européen ISMAP.

Pendant les années 90, au dela de ces mises au point souvent très difficiles et multidisciplinaires, l'Agrochimie devra apprendre comment utiliser ces nouveaux outils pour en tirer le meilleur parti.

Les difficultés de la communication.

J. R. Finney, directeur R&D d' ICI Agrochemicals écrivait récemment : "Historiquement, les résultats obtenus par l'industrie des pesticides ont été importants, on peut même dire remarquables, et ceux qui ont participé devraient être félicités pour leur contribution au bien être de leurs compagnons sur cette terre. Au lieu de cela, leur activité est perçue par la majorité du public comme inutile et dangereuse, à la fois pour l'humanité et pour l'environnement".

Il est certain que l'Agrochimie n'échappe pas au discrédit presque général dont est frappé la chimie. Ce problème de communication s'explique par la conjonction d'un certain nombre de facteurs :

L'agrochimie est une science complexe :
Elle utilise des concepts et un vocabulaire inconnus du public (DL50, ppm, ppb, ppt, NOEL, DJA, demie-vie, cinétique...)
Elle n'apparait dans les médias que lors d'accidents.

Mais surtout le public :
Ignore ce qu'est la chimie (s'imagine qu'on peut à volonté trouver le produit idéal)
Ignore ce qu'est l'agriculture.
Ignore à quoi servent les produits phytosanitaires.
Croit que les coquelicots, les bleuets et les marguerites améliorent les champs de blé.
Croit qu'il existe d'autres méthodes de lutte (lutte biologique).
Ignore la législation qui le protège.
Confond la nécessité de prendre des précautions au moment du traitement avec un risque pour le consommateur de la récolte.
Croit que tous les résidus sont dangereux quel que soit leur niveau.
Généralise ce qu'il sait (ou croit savoir) d'un produit, à l'ensemble des produits (persistance, bioaccumulation, toxicité...)

Certains problèmes relèvent plus d'une attitude psycho-sociologique, que d'un manque de formation ou d'information :
A partir du moment ou ils ont dépassé un certain niveau de satisfaction de leurs besoins, beaucoup de gens refusent de prendre un risque, même infime, s'ils n'en comprennent pas la finalité (zéro risque).
Beaucoup de gens refusent toute espèce de risque qu'ils n'ont pas choisis eux même (zéro résidus).
Incapable de faire la différence entre les arguments de divers groupes de pression le public a perdu une partie de sa confiance dans la communauté scientifique (Les décisions de retrait du marché qui ont été prises à propos du daminozide et la

façon dont elles ont été prises sous la pression de l'opinion, sont pour l'ensemble des scientifiques l'exemple de ce qui ne doit pas être fait).

On voit qu'à l'avenir, l'agrochimie devra faire un effort considérable d'explication et de vulgarisation pour bien faire comprendre les enjeux, la clarté de sa démarche, la qualité de ses études, la fiabilité de ses conclusions, la légitimité de ses choix passés, présents et à venir.

L'élargissement de la notion de responsabilité.

La responsabilité des industriels ne se borne plus, aujourd'hui, à la seule livraison d'un produit loyal et marchand, satisfaisant aux exigences de l'autorisation préalable. Le public exige que les industriels de l'agrochimie aillent souvent au-delà, et prennent en compte tout ce qui concerne un produit, depuis sa fabrication jusqu'à son ultime dégradation dans l'environnement et à la destruction des emballages. Partant du principe qu'il vaut mieux prévenir que guérir, l'agrochimie a dû reconsidérer la notion de "product stewardship", difficilement traduite en français par "éthique et environnement", et créer une fonction correspondante pour assumer des rôles aux contours encore mal définis:
- formation du réseau de distribution et des utilisateurs.
- information des collectivités et du public
- sensibilisation des utilisateurs au respect des bonnes pratiques
  agricoles
- mise à disposition des moyens de protection appropriés
- contrôle de la pertinence des recommandations, etc.

D'autre part, l'évolution permanente des réglementations dans le sens d'une sévérité accrue crée des situations difficiles lorsque de nouvelles normes remettent en cause ou rendent caduques des autorisations obtenues avant leur mise en application.

CONCLUSIONS

La population mondiale poursuit sa croissance à un rythme voisin de 200.000 personnes par jour. Nous sommes donc obligés d'augmenter les récoltes et d'en améliorer la qualité et la conservation. L'agrochimie dans ce domaine joue un rôle irremplaçable en assurant :
1-   La protection contre les ravageurs des plantes (insectes, acariens, nématodes...) destructeurs directs ou indirects des récoltes (notamment en permettant les attaques secondaires de champignons et de bactéries).
2-   La protection contre les maladies des plantes (fontes de semis, pourritures, moisissures, rouilles, mildiou, oïdium, fumagines...) qui diminuent les rendements et abaissent la qualité des récoltes jusqu'à les rendre impropres à la consommation (toxines teratogènes ou cancérogènes comme les patulines et les aflatoxines produites par certaines moisissures, dérivés de l'acide lysergique de l'ergot du seigle...).
3-   La destruction selective des adventices des cultures (mauvaises herbes) qui prélèvent une part importante des aliments et de l'eau nécessaire aux cultures, et qui du fait d'une croissance plus rapide peuvent entrainer la disparition totale de la récolte.
4-   La régularisation des récoltes en empêchant la verse des céréales, en permettant la programmation de la maturation (ananas), en assurant la régularité du calibrage (pommes) etc...

Pour cela l'Agrochimie a su s'adapter en permanence aux besoins des agriculteurs (évolution des cultures et des techniques culturales), en tenant compte de l'environnement et de la sécurité des consommateurs.

Cette adaptabilité est en partie à l'origine des problèmes rencontrés aujourd'hui. D'une part, dans les pays développés, la peur de manquer de nourriture s'est estompée depuis plusieurs générations et le grand public ne se sent plus concerné par les difficultés que connaissent les agriculteurs. D'autre part, il y a un fossé croissant entre ce que savent les rares scientifiques chargés d'évaluer les risques présentés par l'emploi des produits phytosanitaires, et la perception du danger par le public. La mise au point de méthodes analytiques extrèmement sensibles avec des limites de détection souvent inférieures au microgramme par kg en permettant la mise en évidence de traces dépourvues de signification toxicologique n'a fait qu'ajouter à la confusion.

A la fin des années 90, le nombre des sociétées agrochimiques capables de survivre et de maintenir leur potentiel de recherche et de développement ne devrait pas dépasser la dizaine. Ces sociétés devront assumer la poursuite des découvertes de nouvelles structures actives à des doses très faibles et compatibles avec les exigences accrues du respect de l'environnement. Elles devront faire un effort considérable dans le domaine de la communication afin que les mesures réglementaires prises aux niveaux nationaux et internationaux restent guidées par la connaissance scientifique et non par une symbolique archaïque. Elles devront enfin assurer le suivi de leurs produits à tous les niveaux depuis la fabrication jusqu'à l'application, notamment en participant de plus en plus à la formation des utilisateurs.

REFERENCES

R. Delorme et L. Dacol, La Défense des Végétaux, 1988, 249-250, 4.
ACTA, 'Index Phytosanitaire', ACTA, Paris, 1991.
M. Tissut et F. Severin, 'Plantes Herbicides et Desherbage', ACTA, Paris, 1984.
BCPC, 'The Pesticide Manual', Worthing & Hance, London, 1991.
R. Delorme, La Défense des Végétaux, 1989, 259, 26.
GIFAP, 'Toxicological evaluation of pesticides in drinking water', GIFAP, Bruxelles, 1987.
B. N. Ames et L. S. Gold, L'Actualité Chimique, 1991, Novembre-Décembre, 391.

Modern Improvements Brought About by the Chemical Industry: the Direct Oxidation of Olefins, a Contribution to Its History

Fred Aftalion

TOTAL CHIMIE, 24 COURS MICHELET, 92069 PARIS, FRANCE

The direct oxidation of olefins involves the use of air and preferably of oxygen. It seems therefore appropriate to study its historic development in the frame of our Conference.

For over sixty years now, chemists and chemical engineers have worked on this topic since the products derived from such reactions have found many important applications.

However the direct oxidation of olefins does not occur spontaneously and a great deal of research has been necessary before the scaling up of the processes could be made possible.

It is the purpose of this paper to describe the efforts that led to these accomplishments.

Ethylene Oxide

Ethylene glycol had been supplied since 1927 by Union Carbide as an antifreeze for car engines and new uses for ethylene oxide had been found (Di and Triethylene Glycols, Ethanolamines, Glycol Ethers, Ethoxylates...).

However direct oxidation of ethylene was difficult to perform and producers were using the Chlorhydrine route. This required in turn a source of Chlorine and the Chlorine atoms were lost during the reaction in the form of Calcium Chloride :

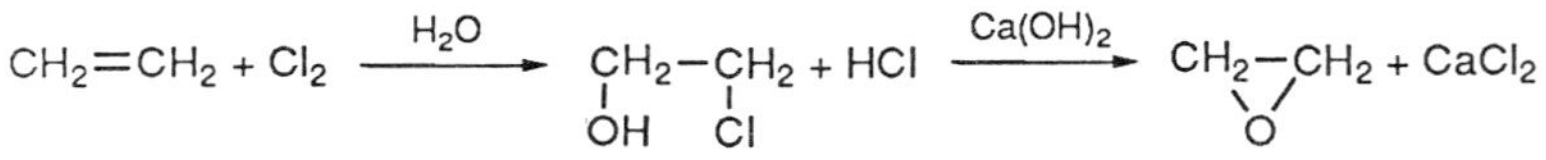

$$CH_2{=}CH_2 + Cl_2 \xrightarrow{H_2O} \underset{\underset{OH \quad Cl}{|\quad\ |}}{CH_2{-}CH_2} + HCl \xrightarrow{Ca(OH)_2} \underset{\underset{O}{\backslash/}}{CH_2{-}CH_2} + CaCl_2$$

Figure 1

A direct oxidation process using a silver catalyst had been patented in 1931 by a Frenchman, T.E. Lefort, but the first industrial unit based on this discovery was built only in 1937 by Union Carbide.

However contrary to what they did later with their Unipol process for Polyolefins, the management of Union Carbide at the time had decided not to licence their technology on Ethylene Oxide.
This prompted two MIT engineers, Ralph Landau and Harry Rehnberg, who had created Scientific Design (SD) in 1946 to study the reaction inasmuch as by 1950 the US consumption of Ethylene Oxide was to reach 200.000 T out of which 70 % were used for antifreeze applications.

Based on the Lefort patent, the study was undertaken in a lab installed in the Manhattan office of SD in New-York, a procedure which would be unthinkable to-day in view of the explosive nature of the reaction.
A fixed bed process was thus elaborated with a yield superior to the one described by Union Carbide.
A pilot plant was now needed to scale up the process. This was provided by Petrochemicals Ltd, a young British company established after the war by Austrian refugees who received in exchange rights to the SD technology.
A more serious difficulty appeared when the time came to find a company willing to carry on the direct oxidation, developed by Landau and Rehnberg, to the industrial scale.
As American chemical companies proved hesitant a recently formed French joint venture between BP, Pechiney and Kuhlmann, called Naphtachimie, accepted to take the risk.
In 1953 the first unit ever built in Europe to produce Ethylene Oxide by direct oxidation of Ethylene started its operation at Lavera near Marseilles.
Having thus proven the feasibility of their process, Landau and Rehnberg were able to sell their technology in the US to Allied Chemical and in 1958 to General Anilin and Film (GAF).
In the mid fifties, Shell which had taken over Petrochemicals Ltd improved the process by replacing air with oxygen which reduced the proportion of residual gases and by the same token minimized ethylene losses.
In 1959 a Japanese company specializing in oxidation catalysts, Nippon Shokubaï was able to further improve the selectivity of the reaction and units in Kazan (Russia), Pyongyang (North Korea), Stenungsund (Sweden) have since been licensed under its technology.
By 1975 already, 60 % of Ethylene Oxide world production, which had reached 6 million tons was using oxygen as the oxidizing agent for ethylene and the capacities of the units from 40.000 T initially were reaching 400.000 T on average.
To-day there are no longer any plants producing Ethylene Oxide through the Chlorhydrine process.
Thus a French discovery was first scaled up by a large US chemical company and later on further improved by young engineering firms in the US and Japan who licensed their technology throughout the World.

Propylene Oxide

Propylene Oxide has become an important raw material for the production of Propylene Glycols, Ethers and Amines.

Polyols derived from Propylene Oxide are used on a big scale to make Polyurethane foams.

Furthermore derivatives obtained by propoxylations such as Glycol Ethers are less toxic that the ones based on Ethylene Oxide.

For a long time the Chlorhydrine route was used exclusively for the preparation of Propylene Oxide inasmuch as the units used for this purpose were no longer needed for the production of Ethylene Oxide and could easily be retrofitted.

Thus a company like Dow who is also basic in Chlorine has remained a prominent supplier of Propylene Oxide through the Chlorhydrin process.

Ralph Landau through Halcon SD and Atlantic Richfield (ARCO) had done, independently from one another, some research work on the direct oxidation of Propylene and had created a 50-50 subsidiary, Oxirane, in order to develop jointly the result of their research.

In 1978 Oxirane had commissionned in Texas a unit to produce directly Ethylene Glycol from ethylene without having to go through the Ethylene Oxide step.

However the process involving Acetic Acid proved very corrosive for the equipment and the unit had to be stopped. Halcon SD then decided to sell his share of Oxirane to Arco who thus became the sole owner of the Propylene Oxide technology.

This technology also studied by Shell is based on the peroxidation with air or oxygen of either Ethylbenzene or Isobutane. The Hydroperoxide obtained in this first step is then reacted with propylene and Propylene Oxide is produced with either Styrene or Tertiary Butyl Alcohol (TBA) as a coproduct :

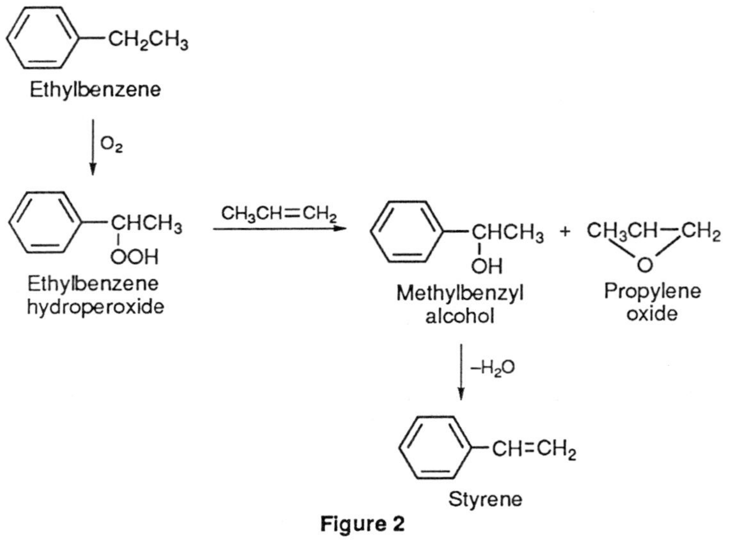

Figure 2

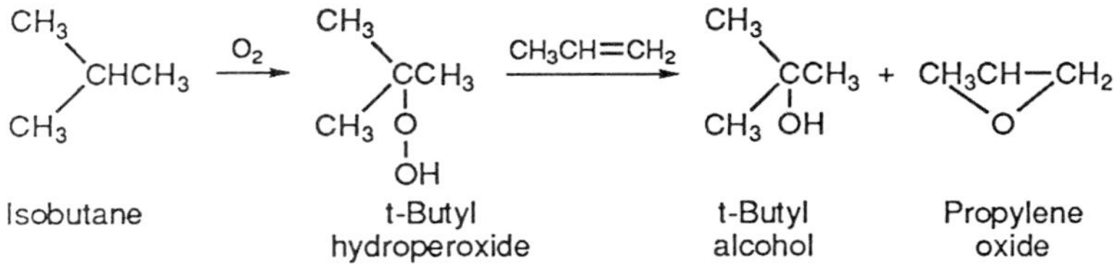

Figure 3

TBA can be used as such as a deicing agent for fuel or in order to boost the octane number of gasoline.
TBA also can yield by dehydrogenation Isobutylene which is reacted with Methanol to produce Methyl Tertiary Butyl Ether (MTBE) :

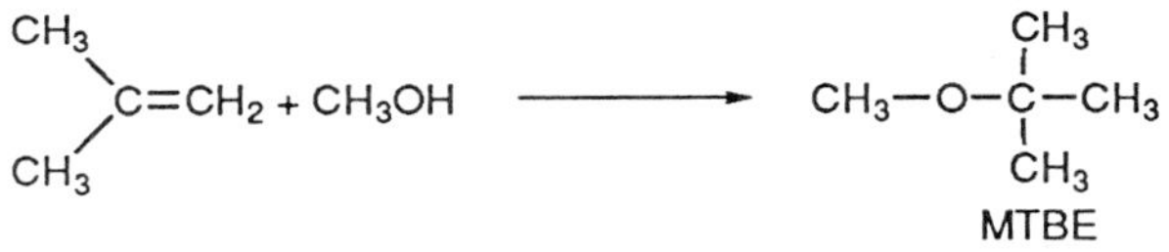

Figure 4

With the implementation of the Clean Air Act in the United States and the progressive withdrawal of Tetraethyl lead in gasoline in Europe, MTBE has become an important additive for lead free gasoline.
Having started a Propylene Oxide production in Bayport, Texas, in 1969 based on the Oxirane process, Arco Chemical has become a world leader for this product, with a 3.3 billion lbs capacity and plants in Holland, Spain, France (Fos/Mer) and Japan. Such units can reach capacities of 400 million lbs each.
Initially the problem was to get rid of a part of the TBA produced by transforming it into Isobutylene or sending it back to the gasoline pool.
In the early 80's when Oxirane became fully owned by Arco Chemical, part of the TBA was mixed with Methanol as an additive for gasoline and part used to produce MTBE.
Starting in 1986, ARCO was able to use all the TBA made at Channelview, Texas, to make MTBE.
Nowadays MTBE has become the main product and there are occasionally difficulties in placing all the quantities of Propylene Oxide that can be obtained at the plant.

Indeed, when Caustic Soda is in heavy demand, Chlorine becomes plethoric and it is very economical to use it to produce Propylene Oxide through the Chlorhydrine process in units that have been totally depreciated.

Since for each two kilos of TBA produced one obtains one kilo of Propylene Oxide, one can readily see that there is a limit to the quantities of Propylene Oxide that can be sold profitably.

In spite of this limitation, Arco Chemical is the first world producer of TBA with a capacity of 5.6 billion lbs and of MTBE with 78.500 barrels/day.

Thus a study started in the 60's by two US companies, the oxidation of propylene has led one of them to a leading position not only for Propylene Oxide but also for MTBE a product that had no outlet at the time the study was started.

Oxidation of Butenes

Methylmethacrylate (MMA) is an important monomer which by polymerisation gives a widely used transparent material known in the trade as Plexiglas or Lucite.

The orignal process developed for MMA in 1934 by ICI and used shortly thereafter by DuPont in the USA and by Röhm & Haas in Germany was based on the reaction of Acetone Cyanohydrin over Methanol in the presence of Sulfuric Acid :

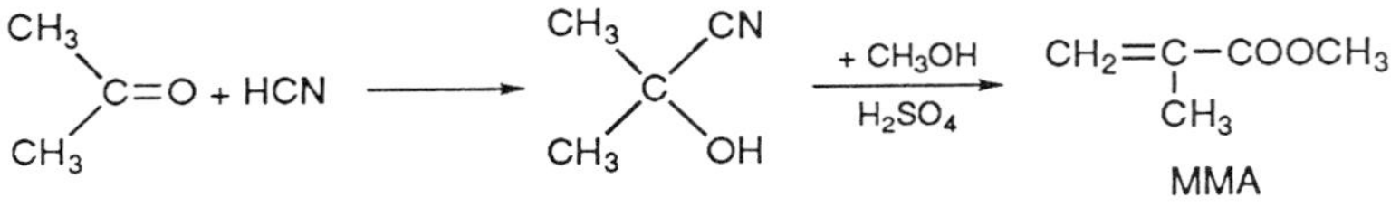

Figure 5

This process still applied to-day in every part of the world has the shortcoming of yielding 2,5 T of Ammonium Bisulfate effluent for each ton of MMA produced.

Its popularity resides in the fact that it is generally practised in units that are already totally depreciated and that the HCN necessary for the production of Acetone cyanohydrin is generally obtained as a by-product in the ammoxidation reaction by which Acrylonitrile is made according to the Sohio (BP) process.

However in Japan a substantial part of the Acrylic fibers used is imported and the installed capacities for Acrylonitrile are not that great, the same being true therefore for HCN.

Moreover the C_4 cuts (BB fractions) are abundant in Japan where 1.8 million tons of C_4 gases are actually burned as fuel each year in the refineries.

This explains why three different Japanese companies have devoted their interest to the oxidation of these fractions with a molybdenum catalyst.

Indeed Mitsubishi Rayon, Asahi Chemical and Nippon Shokubai who set up a joint venture with Sumitomo Chemical, Methacryl Monomer have succeeded in mastering this technology for the production of MMA.

If one starts from Isobutylene as in the process developed in 1982 by Nippon Shokubaî, the reaction is as follows :

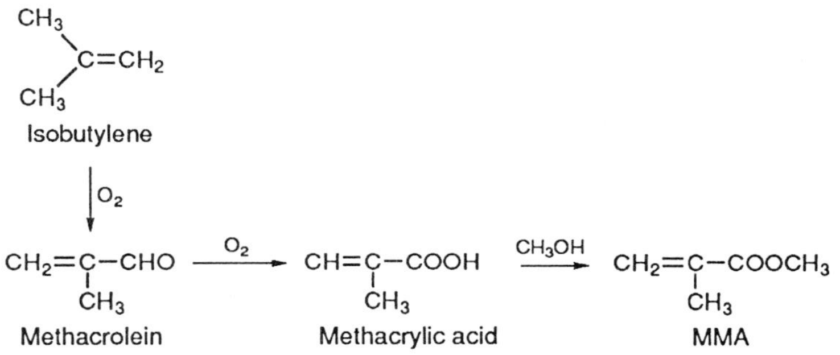

Figure 6

To-day 50 % of the production of MMA in Japan comes from the oxidation of Butenes which makes it possible to upgrade the Butane-Butene fractions from the refineries and avoids the formation of obnoxious effluents.

Acrylic Acid by Oxidation of Propylene

Acrylic Acid is a monomer which finds every day more numerous applications in the form of Esters and Polymers for coatings, thickening agents and superabsorbants in such items as diapers.

The polymerisation of Acrylic Acid was the object in 1901 of a doctorate thesis by Dr. Röhm who was to become the cofounder of Röhm & Haas in Darmstadt prior to World War I.

In 1920 with the help of Walter Bauer, Dr. Röhm had developed a method for producing Acrylic Acid from Ethylene Cyanohydrin.

Later on the chemist Reppe working for BASF succeeded in making the monomer on an industrial scale through the hydrocarbonylation of Acetylene in the presence of $Ni(CO)_4$:

$$CH{\equiv}CH + CO/H_2O \xrightarrow{\text{Ni(CO)}_4} CH_2{=}CH{-}COOH$$

Figure 7

Acetylene however is less and less available to-day and its production involves the use of electric energy which can prove costly in many countries.
As the world production of Acrylic Acid had already reached 550,000T by 1973 out of which 220,000 T were made in the USA, alternate routes for making this monomer had been considered since the late 60's.
One of them was based on the hydrolysis of Acrylonitrile but here again the production of an effluent, Ammonium Sulfate, created a problem.
The most elegant method, also the cleanest, was the one involving the oxidation of propylene with air or oxygen.
This method developed in particular by Nippon Shokubaî in 1970 entails the formation of Acrolein as an intermediary step which is then oxidised in a second reactor using Molybdenum Oxide as a catalyst.

$$CH{=}CHCH_3 + O_2 \xrightarrow{\text{catalyst}} \underset{\text{Acrolein}}{CH_2{=}CHCHO} \xrightarrow{O_2} CH_2{=}CHCOOH$$

Figure 8

It can be predicted that in the future all new units to be erected will use this type of technology which is just as selective as the Reppe method and yields no by-products.

Conclusion

These various examples of basic chemicals obtained by direct oxidation of olefins show how progress is made in our industry.
One starts with processes which, though they need to be improved, allow the production in large quantities of derivatives having already found important applications.
These processes are then refined or modified and on occasions new routes are being devised calling for the use of cheaper raw materials or avoiding the production of obnoxious products.
Sometimes as we have seen in the case of Propylene Oxide the by-product becomes a coproduct or even the main product of the reaction.
This constant evolution of the technology and of the markets offers opportunities for specialised engineering firms which manage to prosper by selling their innovations to the giants of the chemical industry.
Such opportunities have been seized upon in particular by Japanese firms who, though late comers in the field of organic chemicals, have managed to innovate by developing new routes for the synthesis of large volume derivatives.
It is refreshing to think that in our industry the road to prosperity is open to so many different kinds of candidates and that Oxygen plays such an essential function in improving a great number of processes and thereby in lowering the cost of production of chemicals essential to our everyday life.

Taking the Waters: Chemistry and Domestic Water Supply in Victorian Britain

C. A. Russell
DEPARTMENT OF HISTORY OF SCIENCE AND TECHNOLOGY, THE OPEN
UNIVERSITY, WALTON HALL, MILTON KEYNES MK7 6AA, UK

1   INTRODUCTION

A form of pollution that had nothing to do with the chemical industry was
that associated with urbanisation and population growth in the 19th
century. Especially was this true of London. Between 1800 and 1900 the
population of that city rose from 1.1 m to 6.6 m. (the overall population
increased by a factor of only 3). With this increase in population density
came a far greater vulnerability to epidemics of all kinds, including the
dreaded new Asiatic cholera first seen in Britain in 1831.

Incriminated in many of these epidemics, though in a way far from
clearly understood at first, was the public water supply, whether from
shallow wells, springs or an intermittent supply piped from one of the
new Water Companies. And just as guilty in the eyes of many were the
appalling methods of disposing of human waste, especially as much of
this was obviously getting into the water supply, either by seepage or by
the Companies' practice of drawing their water from rivers that were
highly polluted by sewage.

In the case of London's water supply[1] it was impossible to ignore
the potential hazard of the capital's sewage. By mid-century it amounted
to about 31bn gallons per annum, pouring into the Thames through the
sewers in a stream from 2 to 5 ft. high:

> `Through these secret channels rolled the refuse of London, in a
> black, murky flood, here and there changing its temperature and
> its colour, as chemical dye-works, sugar-bakers, tallow-melters,
> and slaughterers added their tributary streams to this pestiferous
> rolling river'.[2]

The urgent need for reform was brought home to the public by a
long series of major epidemics, particularly cholera, typhus and typhoid.
There was also the `Great Stink' of 1858 in which Michael Faraday

dropped his card into the Thames and before it was half submerged the bottom part was rendered invisible in the soup-like water. And there were also certain literary bombshells exploding (about which more later). Some of the most important events which triggered reform are summarised in Table 1. The most important were the outbreaks of cholera in 1831 and 1848.

Table 1 Triggers to reform

```
1828  `Dolphin' scandal
1829
1830
1831  ccccccccccccccccccccccccccccccccc
1832
1833
1834
1835
1836
1837  tttttttttttttttttttt
1838
1839
1840
1841
1842  Chadwick's Sanitary Condition of the Labouring Population of GB
1843
1844
1845
1846
1847  ttttttttttttttttt
1848  ccccccccccccccccccccccccccccccccccccccccccccccccccc
      cccccccccccccccccccccc
1849
1850  Hassall's Microscopical Examination [of London water]
1851  Hofmann's `Chemical Report' on London water [J. Chem. Soc.]
1852
1853  ccccccccccccccccccccccccccccccc
1854
1855  ttttttttttttttttttttttttt
1856
1857
1858  London's `Great Stink'
1859
1860
1861
1862
1863
1864
1865
1866  cccccccccccccccc
1867
1869  tttt
1870
```

          c = 1000 cholera deaths in GB
          t = 1000 typhoid deaths in England & Wales (not distinguished
          from typhus until 1869)

Gradually Victorian England responded to the urgent need, thanks to the work of reformers (like Chadwick), engineers (Rawlinson, Bazalgette) and medical men (Snow). What part did chemists play? The story is almost universally misunderstood and full of rich ironies. There had been a long history of chemical analysis of water: wells, springs and spas. Now chemists were to be at the cutting edge of the new Sanitary Movement and in this case to be associated with removing, not causing, pollution. Their work has recently been documented by Christopher Hamlin in his excellent study <u>A Science of Impurity: water analysis in nineteenth century Britain</u>,[3] though I do not always draw the same conclusions. From (say) 1820 to 1920 chemistry in England had 5 rôles to play in assaulting the problems of massive water pollution.

1 Supportive rôle

That is, a supportive rôle to the evidence of common-sense. In 1827 a pamphlet by John Wright, a journalist, complained of `the Dolphin', the intake pipe of the Grand Junction Co. adjacent to outlet of the Ranelagh, a large, old sewer. It was offensive to eye and nose, and had been blamed by medical men for disease in the area. Common sense was outraged.

Much public outcry followed publication of the `Dolphin' pamphlet. One practical consequence was that sand filters were introduced in 1828 in order to improve the water's appearance (by James Simpson, engineer to the Chelsea Water Co.). Another consequence was the creation of a Royal Commission on Water Supply 1828, consisting of William Brande (chemistry, Royal Institution), Peter Roget (physiology, Royal Institution) and Thomas Telford (engineer). The Government wanted a technical fix. They received and compared many analyses furnished by Companies, but found little consistency. What were the chemists to look for? They might seek `organic matter in a state of decomposition' or H_2S [Lambe], or large amounts of dissolved salts or quantities of sediment [Pearson and Gardner].

John Bostock (physiologist from Guy's Hospital) was commissioned to resolve the problem; he took 49 samples, but rejected many on grounds of taste, smell, appearance and occurrence of micro-organisms. Only two were subjected to quantitative chemical analysis (the best & worst), concluding great variation. Bostock arrived at his main conclusions `without doing any chemistry at all'; chemists then `were in principle free to match any conclusion to any analytical result'.[4]

The problem was uncertainty over the exact causes of disease (though there was plenty of suspicion that water was somehow implicated). `Constitutional medicine' saw disease as imbalance, often due to habits, and postulated a *variety of causes*, rarely a single one. There were various `miasmic' theories of disease which saw its origins in bad air associated with marshland, sewers etc.

Until about the 1850s chemical analysis was used primarily to boost the conclusions of common-sense, to illumine ordinary observations about foul and obviously polluted water, and to act as a useful supplementary indicator of the value of any individual source of supply.

2 Contentious rôle

In arguments about water supplies chemistry entered a dangerously new and contentious phase. The 1850s were ushered in by the worst cholera epidemic ever to strike Britain, in 1848/9 (62k deaths). There was fierce debate for the next 3 years, and some action. In London those Companies drawing water from the Thames were forced (a) to move upstream for their intake (so avoiding the worst of the sewage contamination, but still receiving that from 800k people even further upstream), and (b) to filter their water.

The mounting public indignation was fuelled by Chadwick's return to the fray with two volumes for the General Board of Health on the Metropolis Water Supply, and by Hassall's <u>Microscopical</u> <u>Examination of the Water supplied to the Inhabitants of London and Suburban Districts</u> (both 1850). The microbial life depicted was meant to horrify, but what was exactly was its significance? In 1849 Snow identified water supplies as a cause of cholera, and William Budd announced the discovery of a water-borne cholera fungus. But the implications were far from clear, even to those who accepted these fimdings. Granted even a general connection between germs and disease, one could adopt one of three positions concerning microbes. They were:

1.   *Direct cause of cholera etc.*: few held this apart from Budd.

2.   *Indicators of cholera etc.*: thus Hassall believed the cause to lie in the decomposing organic matter on which the microbes fed.

3.   *Simply disgusting*: this was a powerful argument, emotionally if not logically. Angus Smith described those microbes in the Thames as `larger, fatter and uglier than any preceding'. Aesthetic considerations were never far away.

Chemists as well as microscopists entered the fray, though chemists were generally more successful, partly because they were more experienced in public argument and partly because few agreed that Budd had proved his point. Chemists were enlisted by all sides, by the Companies who were fighting against proposals to end their monopoly, by reformers who wanted to do just that, and by a harassed government (see Table 2). It is interesting to note how several chemists served more than one party, often at the same time or within quite short intervals.

Most of the chemists acted for the Companies, and tended to look for large concentrations of H_2S, PH_3 etc. These never turned out large enough to cause concern. Hofmann *et al*. sought to measure the concentration of organic nitrogen as an index of the water's potential for putrefaction, hoping to do this by subtracting the value for inorganic nitrogen from the total nitrogen concentration.[5] The results were ambiguous, and controversy continued till the end of the century.

3 Diagnostic rôle

In the last third of the 19th century chemistry sought to assume a diagnostic rôle in the assessment of water quality. In 1866 cholera struck again. Suggestions were now made to supply London from Wales or even the Lake District.

<u>Table 2</u>   Where were the chemists?

	Reformists	Water Companies	Government Commissions
1820s, '30s & '40s		A Aikin	*RC Met. Water Supply 1828*
		W T Brande R Phillips	W T Brande
1850s	*For Chadwick*	J T Cooper	`Chemical Report ' 1851*
	R A Smith L Playfair A W Hofmann *For alternative suppliers* D Campbell J Stenhouse T Clark W A Miller	A S Taylor	W A Miller T Graham A W Hofmann *Gen. Bd. Health Rept. 1855* L Blyth A W Hofmann *RC Sewage of Towns 1857-61* J B Lawes J T Way
1860s & '70s	E Frankland B Brodie	H Letheby	*RC Rivers Pollution 1865-8* J T Way *RC Water Supply 1867-9* E Frankland W Odling *RC Rivers Pollution 1868-74* E Frankland
1880s & '90s	E Frankland	C M Tidy J Dewar W Crookes	*RC Met. Water Supply 1892-3*

At this point one man emerges who was to dominate this and many other aspects of British chemistry for almost the rest of the century: Edward Frankland (1825-1899). One year previously he had succeeded A. W. Hofmann as Professor of Chemistry at the Royal College of Chemistry (now RSM). With this job went the relative sinecure of monthly analyses of water supplied by the eight London Companies.[6] In 1868 he was appointed to the new Rivers Pollution Commission: a 6-year task at £800 p.a. He was the only chemist on the Commission, which soon became his mouthpiece. He performed over 5000 analyses for the Commission, as well as continuing his monthly reports for the Local Government School[7]; he retained a quasi-official position as `water analyst' till 1899, giving him an enormous influence, at home and overseas. His analyses were in demand from India, Mecca, China and many other places.

The methods of water analysis were quickly improved. At first organic matter had been estimated by loss of weight on ignition, but this was soon abandoned as inorganic matter (as nitrates) also decomposed under those conditions. In 1867 Wanklyn *et al.* announced their process on the basis that the lethal organic constituents were albuminoid.[8] The water was first heated with alkali to expel ammonia (from urea, NH_4^+ salts etc.), and then treated with alkaline $KMnO_4$, the ammonia being distilled off into Nessler's solution, and estimated colorimetrically (by eye):

$$\text{albumen + alkaline } KMnO_4 \longrightarrow NH_3$$

Frankland at first used $KMnO_4$ oxidation, titrating it into the water, and assuming the volume added to be a direct measure of the total amount of putrescent material present. He soon became sceptical about Wanklyn's method on the grounds that there is no evidence that polluted water does contain albumens, the nature of the nitrogenous organic substances being `utterly unknown to chemists'.[9]

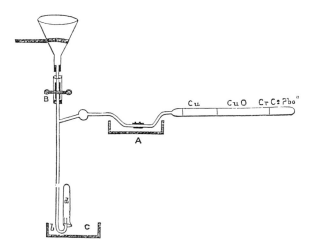

Figure 1: Frankland's combustion apparatus

By 1868 Frankland had devised a new method for estimating organic C and N (see figure 1).[10] Water was saturated with SO_2 (to prevent further oxidation) and evaporated to dryness. The residue was mixed with PbCrO and transferred to a combustion tube which was then evacuated by a Sprengel pump.

Combustion was commenced (45 - 60 min.) and the resultant gases (CO_2, NO and N_2) were collected by further application of the vacuum pump. Frankland later devised a modified apparatus devised for simplified gas analysis[11]:

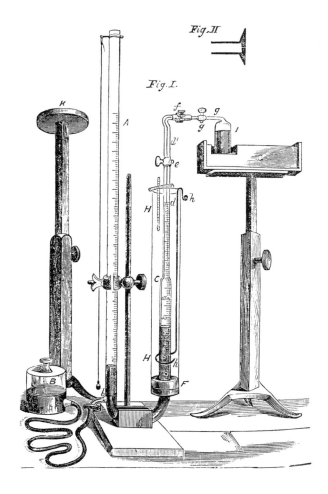

Figure 2: Frankland's apparatus for analysing gases from water analysis

Frankland claimed that 8 years' experience showed this method `the only one yielding quantitative results in any degree trustworthy'.[12]    The Wanklyn-Frankland row became a veritable *cause célèbre*. Wanklyn claimed his rival's method `has not commended itself to the generality of chemists, and has never come into general use'.[13]   Frankland admitted that the ammonia process was

very popular.[14]   In fact, it was much quicker and simpler and did not require
Frankland's superb experimental technique. Nor, as Crookes pointed out, did it
involve such expensive equipment, which made water analysis possible only for
the *élite*.[15]

Correspondence still extant reveals that Frankland's method was used by E.
B. Truman for `hundreds of analyses' in 1874[16] ; also by J. W. Thomas of
Cardiff, who, to keep fees down had temporarily gone over to Wanklyn's
method in 1876 (though hoping to return to Frankland's process as soon as
possible)[17]; and by Odling (though he drew different conclusions).[18]   An
undated memo by Frankland indicates its use by Russell, Hill, Bischoff,
Campbell Brown, H. Brown, O'Sullivan, Moulton, Odling, Williamson,
Truman, Tate, Donkin, Thomas.[19]

Table 3 London's water supply, 1876

Source	Company	Date of origin	Proportion of organic matter
Wells from chalk hills nr.Crayford	*Kent**[*]	1809	1.0
R.Lea (Essex)	*New River*	1619	0.9
R.Lea (Essex)	*E. London*	1807	2.4
R. Thames	*W. Middlesex*	1806	2.8
R. Thames	*Grand Junction*	1811	3.3
R. Thames	*Lambeth**[*]	1785	4.1
R. Thames	*Chelsea*	1723	4.2
R. Thames	*Southwark**[*]	1834	4.5

[* = south of the Thames]

What was Frankland measuring? Not a specific toxin but a measure of
previous sewage contamination. `It cannot be too widely known that chemical
analysis is utterly powerless to detect any matter positively injurious to health in
any of the forms of animal refuse which go to contaminate water. The use of
chemical analysis in the investigation of water for sanitary purposes lies almost
exclusively in a different direction. It is only by bringing to light the previous
history of potable water that chemistry can help the custodians of public
health'.[20]

As the years went by Frankland came ever closer to attributing the causes
of water-borne diseases to micro-organisms. Thus this chemical analysis was
ultimately concerned, not with the *composition*, but with the *history* of a
sample. Even if sewage itself be removed by percolation through the soil, a
`skeleton of the sewage' remains in form of nitrates and nitrites, and they may

indicate the otherwise undetectable causes of disease. So the estimation of
inorganic nitrogen (nitrates, nitrites etc.) became by the early 1870s of equal
importance to the determination of organic matter. The reception of his work
was varied.

That chemical analysis cannot discover the ultimate cause of cholera was
accepted by Miller, Brodie, Angus Smith and William Odling. However not all
agreed with Frankland's fully enunciated doctrine of previous sewage
contamination. Other chemists, especially those employed by the water
companies, made light of Frankland's assumption and even with a high organic
and/or nitrogen content declared their waters entirely fit to drink. And, of
course, it was also possible to denigrate Frankland's method, saying its results
were always too high.

On the medical side Frankland was severely criticised by Letheby (MO for
the City of London), and especially by C. M. Tidy (MO for Islington), who
opposed him on almost every conceivable issue relating to water supply. His
chemical opponents included (as well as the permanently implacable Wanklyn)
Odling, Crookes, Dewar and others like Letheby and Tidy `willing to tailor their
statements to the companies' needs'[21]. In preparation for `a great discussion at
the Chemical Society' Frankland sought moral support from his friend John
Tyndall.[22]   The question was:

> whether running water can be safely used for dietetic purposes a
> few hours after it has been mixed with sewage. Dr. Tidy
> contends that it can, on the ground that the dead organic matter
> of the sewage will be burnt up by spontaneous oxidation, and the
> living germs will burst their envelopes by endosmic action and
> die!

He asked Tyndall to say `even a dozen words about germs', in which case
`we should not leave Tidy a leg to stand upon'. To which his friend remarked
that:

> The levity with which certain of our medical men throw off their
> hypotheses regarding germs is extraordinary. They shut their
> eyes to the most conclusive evidence, and give themselves to the
> strangest delusions'.[23]

In the event Tidy's theories were savaged by Huxley, and victory for
Frankland was `complete and crushing'.[24]   Such disputes were not primarily
about chemical results but rather their interpretation. With hindsight we may
see that Frankland's instinct was mainly correct, but his greatest service to
Britain was made by using chemistry in a different way.

4 Rhetorical rôle

It will now be clear that there often was more than one way of interpreting
a given set of analysis results,- hence much of the controversy. Frankland
himself made full use of this ambiguity. In some cases it is hard to avoid the
impression that his interpretation was conditioned by other, non-chemical
considerations, most notably by his own understanding of the dangers of
sewage contamination. Chemistry thus complemented other information.
Wherever he could Frankland used high organic/ nitrogen content to condemn
the water supplied by water companies. As he said, `my motto, unlike that in

criminal cases, has always been "assume water to be guilty until it is proved innocent"'.[25]

Relentlessly he argued for nearly 30 years for Government action and for the Water Companies' reform, even if that meant their municipalisation. He was aided by two related things: the astronomically large number of water analyses emanating from his own laboratory, and his own towering reputation as the leading water analyst of Britain. Hamlin has argued that `he shamelessly used his position as a forum for social change'.[26]

Chemically he had few new facts to argue from but for three decades he extracted from his analytical results every possible ounce of political capital, urging closure of water supplies here, drawing water from deep springs or wells there, and rigorously monitoring it all by the very best chemical means available. As he wrote late in life: `Take the public into your confidence'.[27]

Slowly the government did respond to public pressure. In 1871 the Metropolis Water Act created the post of Water Examiner submitting monthly and annual reports to the Board of Trade (from 1872 to the LGB). The Public Health Act of 1875 decreed that a Medical Officer of Health must be employed by sanitary authorities, and a Public Analyst by municipal authorities. The subsequent increase in visibility and controversy led to the foundation in 1877 of both the Society of Public Analysts[28] and the Institute of Chemistry[29], the one particularly associated with Wanklyn and the other with Frankland. It is impossible to exaggerate the influence of the chemical rhetoric on these events. However efforts to centralise control of the London Water Companies did not succeed until the establishment of the Metropolitan Water Board under the Metropolis Water Act of 1902, three years after Frankland's death.

Of course this rhetorical use of chemistry made him many enemies. In the subsequent exchanges he does not always emerge unscathed, nor indeed does it seem right that he should. Thus Sir Alexander Binnie, Engineer to the London County Council, alleged that Frankland received 800gn for analytical services leading to his evidence given to the Royal Commission on 15 Dec. 1893. This evidence was inconsistent with that to London County Council in previous June. Binnie suggested Frankland changed his tune after receiving a final payment from the LCC, and was now acting in interest of water companies![30] The accusations were indignantly denied by Frankland[31] and Binnie had to retract some them. But it would not have been the first time that Frankland found himself changing sides in a legal battle.

The closing decade of Frankland's life was marred by a tragic quarrel with his son and former partner, Percy Frankland (ostensibly about fees). In one letter Frankland demands the return of certain blank sheets of paper signed by himself as analysis report forms![32] His son retorted that had this been known to the public `the consequences might have been very serious'![33]

5 Positive rôle

As the years went by Frankland came ever closer to attributing the causes of water-borne diseases to micro-organisms. Here he was assisted - even urged on - by his son Percy Frankland (1858-1946).

With our own knowledge of pathogenic bacteria we can see why (as Frankland said) chemical analysis could never give the last word on water purity. It had to be supplemented by microscopy. Towards the end of his life

Frankland claimed that he had been doing bacteriological work on water supply since 1871.[34]   By now this most distinguished of English chemists was fully committed to a joint attack on the problem by chemistry and microbiology. He was much aggrieved, therefore, to receive a request in 1898 from LGB that he exclude bacteriological results from his monthly report: commissioned not by LGB but by Metropolitan Water Companies.[35]   Frankland objects: `the virulence of pathogenic microbes is due to the chemical poisons they excrete'.[36] With typical Frankland polemic he complains that his reports have been mutilated by LGB, - all the most important results having been omitted.[37]

But it was not a question merely of analysis. Now it became possible not merely to diagnose trouble, but also to prescribe remedies. The germs simply had to be removed.

In May 1892 Frankland began serious bacteriological researches and could no longer deny the beneficial effects of filtration.[38]   Thus many of his previous criticisms were shown to be unnecessarily harsh. In 1878, for example, he had specifically denied that filtration could remove `typhoid poison' from water.[39] The Water Companies were of course delighted. It was discovered that in the top 10-25 mm. of sand filters worms, larvae etc. remove germs and oxidise organic matter. Another procedure to have a similar if unexpected effect was Clark's process for water softening; Frankland speaks of its efficiency in removing microbes, a fact made clear by Percy Frankland.[40]   This, of course, gave an interesting ***post hoc*** justification for Frankland's advocacy in the early 1870s of intermittent filtration to remove organic ***débris*** from sewage (including overseas locations).[41]

During the closing years of the previous century several attempts had been made to destroy germs by chemical means, none of them successfully on a large scale. The solution came at Lincoln in 1904-5 when Alexander Houston, Bacteriologist to the Sewage Commission, was called in to deal with an urgent problem of typhoid bacilli spread by flood waters and causing a severe local epidemic. The remedy was to use chlorination, in the form of ICI's `Chloros' (aq. NaOCl). Shortly afterwards Houston became the Metropolitan Water Company's first Director of Water Examination, and introduced similar methods in London. By 1911 800m gal per day were being treated thus world-wide. Chlorine gas was first used for this purpose in 1910 (Darnell, US Army). Superchlorination was introduced later for very polluted waters (excess chlorine being removed by SO_2, active Charcoal etc.). Ozone was used in France till quite recently.[42]

3   CONCLUSION

Hamlin[43] criticises the view[44] that chemical (and biological) analysis led to clearer insights into conditions needed for public health. He points to :

(a)   The disunity of chemical knowledge - but who said chemists had to agree before useful information emerges? Controversy is the very stuff of science.

(b)   The fact that scientific discoveries did not always lead to technical action (as in case of Snow's recognition of the cause of cholera); but that does not mean that they never did, and one must judge by the facts. Frankland's advocacy of intermittent filtration, although never patented, led to its adoption by many towns, as Kendal and Birmingham.[45] Chlorination is another obvious example.

(c)   The view that knowledge strengthens the mandate for social changes. He observes that this did not always happen. Human nature being what it is one would have hardly thought this surprising. Hamlin could have quoted Frankland's own rueful comment on the public impact of even negative comments. After he had shown the uselessness of a certain method of sewage treatment (the `A.B.C.' process) the shares of the owning company promptly rose by 800%![46] But the very employment of water analysts, the insistence on regular reporting, etc. all add up to a major response to the perceived benefits of chemical analysis.

What of Edward Frankland? For all his faults he was a scientist committed long before his time to reducing pollution, and was deeply concerned with the whole urban environment. His almost paranoid obsession with water purity reflected this concern. In his case it is easy to focus only on analysis of water supplies and overlook his other work for public health, such as detection and estimation of industrial effluents and lead pollution, and research on sewage purification. A measure of his national stature came with the bestowal of a knighthood, in the Queen's Jubilee Honours List in 1897. It was widely reported that this reflected not his discovery of `a new law of nature'[47] [valency] but `in his more ordinary professional capacity as water analyst to the Home Department'.[48]

NOTES

Grateful thanks are expressed to Mr and Mrs Raven Frankland for permission to cite from documents in their possession (the `Raven Frankland archive'); microfilms of these documents are at the Open University.

1   See Anne Hardy, `Water and the search for public health in London in the eighteenth and nineteenth centuries', `Medical History', 1984, 28, 250-82.

2   E. Walford, `Old and New London', Cassell, London, n.d. (c.1880), vol. v, p.238.

3   C. Hamlin, `A Science of Impurity: water analysis in 19th century Britain' (Hilger, Bristol, 1992).

4   Hamlin, `A Science of Impurity', pp. 87, 85.

5   T. Graham, W. A. Miller and A. W. Hofmann, J. Chem. Soc., 1851, 4, 375-413.

6   See C. Hamlin, `Edward Frankland's early career as London's official water analyst, 1865-1876: the context of "previous sewage contamination"', Bull. Hist. Med., 1982, 56, 56-76.

7   The LGB was created in 1871 as the successor to the old Poor Law Board; from 1875 it took over the Registrar General's responsibility for water analysis and other sanitary matters.

8   J. A. Wanklyn, E. T. Chapman and M. H. Smith, J. Chem. Soc., 1867, 20, 445-54.

9   E. Frankland, Royal Commission on Water Supply to the Metropolis and Large Towns: Minutes of Evidence, P.P. 1868-9 xxxiii, p.339.

10   E. Frankland and H. E. Armstrong, J. Chem. Soc., 1868, 20, 77-108.

11   E. Frankland, J. Chem. Soc., 1868, 21, 109.

12   E. Frankland, J. Chem. Soc., 1876, 29, 825.

13   J. A. Wanklyn, `Water-Analysis', Kegan Paul, London, 10th edn., 1896, p.34.
14   E. Frankland, J. Chem. Soc., 1876, 29, 826 (847).
15   W. Crookes, Chem. News, 1873, 28, 37.
16   Letter from E. B. Truman to Frankland, 25 June 1874, Raven Frankland archive, Open University Microfilm no. 01.04.0409; Truman was the Public Analyst for Nottingham.
17   Letter from J. W. Thomas to Frankland, 22 December 1876, Raven Frankland archive, Open University Microfilm no. 01.04.0415.
18   Letter from W. Odling to Frankland, `27 April', Raven Frankland archive, Open University Microfilm no. 01.04.0418.
19   E. Frankland, undated memorandum, Raven Frankland archive, Open University Microfilm no. 01.04.0421.
20   Letter from Frankland to Registrar General, 10 July 1869, Raven Frankland archive, Open University Microfilm no. 01.4.1134-1150 [1147], nearly identical to published letter in Rivers Pollution Prevention Commission, 5th Report, 1874, p.51.
21   Hamlin, `A Science of Impurity', p.191.
22   Letter from E. Frankland to J. Tyndall, 14 May 1880, Royal Institution archives 9/E6 20.
23   Letter from J. Tyndall to Frankland, 16 May 1880 Raven Frankland archive, Open University Microfilm no. 01.04.1661-3.
24   Letter from E. Frankland to J. Tyndall, 21 May 1880, Royal Institution archives 9/E6 21.
25   E. Frankland, undated MS memorandum (c.1898) Raven Frankland archive, Open University Microfilm no. 01.03.0865.
26   Hamlin, `A Science of Impurity', p.153.
27   E. Frankland, undated memorandum, (c.1896) Raven Frankland archive, Open University Microfilm no. 01.04.0427.
28   R. C. Chirnside and J. H. Hamence, The Practising Chemists: a history of the Society for Analytical Chemistry, S.A.C., 1974.
29   C. A. Russell, N. G. Coley and G. K.Roberts, Chemists by Profession: the origins and rise of the Royal Institute of Chemistry, Open University Press, 1977; in 1896 Branch E of the Institute's examination was established: `the analysis of water, foods and drugs'.
30   Letter from A. R. Binnie to E. Frankland, 20 May 1898, Raven Frankland archive, Open University Microfilm no. 01.03.0859.
31   Letter from E. Frankland to A. R. Binnie, 13 May 1898, Raven Frankland archive, Open University Microfilm no. 01.03.0868.
32   Letter from Edward Frankland to Percy Frankland, 29 November 1889, Raven Frankland archive, Open University Microfilm no. 01.04.0166.
33   Letter from Percy Frankland to Edward Frankland, 12 December 1889, Raven Frankland archive, Open University Microfilm no. 01.04.0219.
34   Letter from E. Frankland to Local Government Board, 13 March 1899, Raven Frankland archive, Open University Microfilm no. 01.04.0541.
35   Local Government Board to E. Frankland, 30 November 1898 and 27 February 1899, Raven Frankland archive, Open University Microfilm nos. 01.04.0537 and 0539.
36   Letter from E. Frankland to Local Government Board, 18 October 1898, Raven Frankland archive, Open University Microfilm no. 01.04.0534.
37   E. Frankland to Local Government Board 6 February 1899, Raven Frankland archive, Open University Microfilm no. 01.04.0548.
38   Copy-letter from Frankland to A. Binnie, 28 May 1898, Raven Frankland archive, Open University Microfilm no. 01.03.0844.
39   Report of Select Committee for the Public Health Act (1875) Amendment Bill, P.P. 1878, xviii, p.41

40   E. Frankland to E. Surrey Water Co., 22 March 1898, Raven Frankland archive, Open University Microfilm no. 01.04.0498.
41   E.g. Letter from Frankland to J. F Bateman, 16 November 1874, Raven Frankland archive, Open University Microfilm no. 01.01.1107 (the water was for Constantinople).
42   W. S. Chevalier, London's Water Supply 1903 - 1953, Staples Press, London, 1953, pp.253-70; Anon, Water Sterilisation, ICI, n.d.
43   Hamlin, `A Science of Impurity', pp. 302-2
44   C. A. Russell, Science and Social change 1700-1900, Macmillan, London, 1983, p.257.
45   E. Frankland, Sketches from the life of Sir Edward Frankland, London, 1902, p.145.
46   Ibid, p.146.
47   Proc. Inst. Civil Eng., 1900, 139, 343-49 (349).
48   Nature, 1899, 60, 372.

Priestley Lecture 1992

The Stellar Nucleosynthesis of Oxygen

Hubert Reeves
CENTRE NATIONAL DE RECHERCHES SCIENTIFIQUES

Transcription of Lecture prepared by R. Corns, Société de Chimie Industrielle, 28 rue Saint-Dominique, 75007 Paris, France

The assumption of oxygen as deriving from stellar nucleosynthesis makes it inseparable from the origin of the various other chemical elements and requires therefore by way of introduction a summary overview of the history of nucleosynthesis theory.

Professor Reeves recalled that nucleosynthesis was equated early in the century to Darwinism, chemical elements and their isotopes being considered counterpart to animals and plants with comparable characteristics of variety and abundance. Hence, in both contexts, there was a shared belief that none of the chemical elements or living organisms was fixed for eternity but followed an evolutionary course. It was in the late twenties that nuclear physics was to provide the key to the dynamics of evolution around the idea of nuclear force, i.e. when two nuclei collide, they have a chance of fusing and producing something else. Conversely, to explain why certain nuclei do not come together and fuse, for example water and air, it was advanced that it is merely because their respective nuclei are electrically charged and tend to repel each other, thereby with little chance of colliding. It was a young physicist, Fred Utermans, who reasoned that the only way to overcome such electrical repulsion (e.g. between the nuclei of air and water) was to bring the gas to a very high temperature where the kinetic motion would be strong enough to bring them together and set off nuclear reaction. One of the places where such high temperatures exist is in stars.

Subsequent research was to generalize the idea that all elements could be produced by the thermonuclear reactions taking place in stars. At this point, the speaker illustrated the range of elements from hydrogen to uranium with a corresponding scale of relative abundance. A major characteristic of our universe is that 90% of the atoms are hydrogen and almost 10% helium, the rest being less than one part in a thousand. Among the elements the third most important is actually **oxygen**, more specifically **oxygen-16**. The fact that 90% of the universe is still hydrogen merely shows that the nucleosynthetic process of stars has not gone very far. It has hardly touched 10% of matter which accounts for the 92 or so stable elements known to mankind. Furthermore, the Coulomb slope demonstrates that the heavier the nuclei, the harder it is to get them to touch one another. Consequently, the places in the universe where temperatures are high enough to provoke nuclear reaction tend to become progressively fewer in number.

The speaker explained the phenomenon of nuclear stability or nuclear force, how it is calculated for each element according to its mass, the lower the mass the greater the stability, and the more energy required to break it up or provoke its integration. The element with the highest stability or the lowest mass is iron-56. This means theoretically that any matter whatsoever, e.*g.* from dirt to strawberries, brought to a temperature high enough for all nuclear reactions to be in equilibrium with one another, will be ultimately transformed into iron, the most stable element and consequently the most abundant. In this way, nuclear physics demonstrates the temperatures of billions of degrees and the kinetic energy of millions of volts producing the thermal equilibrium around the iron peak, and thereby provides an insight into the past of astronomy. Hence, stability and abundance are deeply influenced by the electromagnetic and nuclear properties of the elements while they are being formed in the stars.

Professor Reeves went on to comment briefly on the physics of stars and the circumstances of the creation of the nuclear elements. Despite their gaseous state, stars are stable, which accounts for their very long existence, *i.*e. billions of years. Their stability is due to a delicate equilibrium between two forces: **gravitational pull** toward the centre of the star and the **heat pressure** generated by stellar collapse. The extreme heat provokes the flux of photons from the hotter to the cooler region of the star, this release of energy being the physical principle explaining why stars shine.

However, stability is not perfect, and stars are really in quasi static equilibrium, being less and less able to support their own weight. They are in fact characterised by a negative heat capacity according to which, contrary to the general law of physics, the more energy they emit, the hotter they become. The reason for this phenomenon is that stars, which have a strong gravitational energy, behave inversely and instead of becoming cooler with the release of energy they grow hotter. The story of stars is therefore one of a constant increase in heat and density which can be measured over time by their luminosity curve. However, when the temperature of a star becomes high enough for the protons to touch one another, they fuse and transform into helium. At this point the star departs from the gravitational mode of getting its energy and its temperature stops increasing.

Hence, contrary to a widespread belief that nuclear reactions account for the heat of stars, what happens is quite the reverse, the nuclear reactions preventing the star momentarily from becoming hotter. To obtain the energy necessary to remain stable, the star lapses into the gravitational mode which boosts the temperature to a point where the gravitational energy falls with a corresponding rise in the nuclear energy, thus enabling the star to continue to shine, i.e. keep the same radius, mass, temperature and aspect, for as long as it takes to burn this fuel. At the present time, the sun is in this phase, following upon a period in its past during which it collapsed from the state of interstellar gas.

Thus, the life of a star is a sequence of periodic gravitational contraction, followed by nuclear reaction, then again by gravitational contraction which pushes the temperature up again to a point where the ashes of the previous period become the fuel for the following period. For example, hydrogen burns at 10 to 30 million degrees and transforms into helium which cannot however burn at the same temperature as hydrogen, requiring for this some 200 to 300 million degrees. When the star through gravitational contraction reaches this temperature range, a new period of thermal nuclear reaction takes place with the helium becoming the fuel of the star for a certain time. This will continue until all the helium has been transformed into carbon and oxygen.

Reverting to the history of the sun, the latter has been burning hydrogen into helium for the last 10 billion years. In another 5 billion years it will warm up again to approximately a temperature of 200 million degrees and will burn helium into carbon and oxygen for around 300 million years. This period will become increasingly shorter due to the extremely rapid combustion process as well as to the decrease in the binding energy during the transformation from helium to carbon and iron with the burning of its carbon and oxygen into heavier nuclei.

The speaker used the Hertzsprung-Russell diagram to illustrate further the evolution of stars and their internal activity. The plotting of the stars in the sky as a function of their colour or surface temperature reveals a contrast between highly populated regions - called the main sequence - and regions which are almost empty. One such area contains the red giants, like Antares which is easily visible to the naked eye, where helium is being transformed into carbon and oxygen. The super giant region is where later nuclear reactions are taking place with formation of neon, magnesium and all other elements to iron. The Hertzsprung-Russell diagram can thus be seen as a snapshot of the stars with their various ages and masses. It can also be used to reconstitute the history of a given star. For instance, the sun was born as a very diffuse cloud, with low luminosity, becoming a very red cloud of interstellar matter for approximately 10 million years. Deriving its energy from gravitational collapse, after some 25 million years it landed on the main sequence where it was hot enough to undergo nuclear fusion, namely the transformation of hydrogen into helium. It has been in this state for 5 billion years and will remain so for another 5 billion years. Its core hydrogen will be transformed into helium with the result that it will have no more energy and will start to collapse, thereby warming up to become a red giant. It will become very large, its radius extending from its centre to the earth or Mars and causing the volatilisation of these planets. The sun will continue for a time to transform its helium into carbon and oxygen which it will eject on becoming a planetary nebula.

However, depending on their size, all stars do not have the same fate. Small stars on collapsing become rapidly electron degenerate, the degeneracy of the electrons retarding the process of collapse and unlinking the history of the star from its thermal history. As the electrons take care of keeping it stable, the small star does not need to warm up any more with the result that it grows bigger and loses most of its envelope, becoming successively a planetary nebula and a white dwarf, namely a star with no longer an energy source, a degenerate body irradiating away in space as finally a black dwarf.

Big stars have a much more dramatic fate, essentially because they never become electron degenerate. They keep on collapsing, ejecting most of their matter into space in a very large explosion. In this connection, the speaker made reference to the supernova observed in the Magellan Cloud on 23 February 1987 which provided confirmation of the assumption based on nuclear physics that supernovae are actually stars in process of collapse. He commented on the successive stages of the process over time, illustrating his explanations by a series of slides. He characterised the supernova as a celestial chemistry laboratory where all the atoms made in the stars at very high temperatures are ejected into the very low temperatures of space where they undergo electromagnetic reaction. These atoms join together to form most of the molecules in the universe (*e.g.* water, methane, ammonia, carbon dioxide).

One of the most important outcomes of such chemical reactions is dust, and dust is only possible where there is oxygen. And without oxygen there can be no solid bodies like the earth or biological evolution. Hence, one key to the evolution of life is the presence of oxygen ejected from stars and its junction with other metals in order to form solid bodies. The speaker illustrated such manifestations of oxygen by spectroscopic views of the constellation of the Pleiades and of the Northern Lights.

Professor Reeves referred in conclusion to some of the enigmas related to the existence of oxygen in the universe. He gave details of the initially problematic and unlikely theory of resonant reaction advanced by Fred Hoyle in 1948, and confirmed several years ago, as an explanation for the transformation of helium into carbon and eventually oxygen. Regarding such numerous strange coincidences which are nowadays increasingly monitored and the subject of much discussion and debate, the speaker preferred to ,consider them as demonstrating the principle of complexity which in his opinion holds as true in respect of humans, birds, ants and molecules of sugar as it does for the prevalent cosmological model of the Big Bang. For instance, while the universe is considered to be expanding, the general theory of relativity does not specify the way these objects move with respect to one another. Consequently, although there are therefore many modes of expansion which could be adduced in compatibility with Einstein's model, none of them accounts satisfactorily for the complexity of the atoms required to form galaxies.

Chemical Synthesis

Reactor Engineering Aspects of Gas–Liquid Oxidation Processes

J. Breysse
RHÔNE-POULENC INDUSTRIALISATION, 24 AVENUE JEAN JAURÈS,
60153 DECINES CHARPIEU CÉDEX, FRANCE

The term oxidation includes a great variety of chemical processes such as :

. the production of energy by combustion of fuels and mixtures of the same kind

. the selective oxidation which gives access to certain intermediates or fine chemicals

. the slow oxidation of a large number of organic compounds and polymers which corresponds generally to ageing phenomena :

the biological oxidations
and some types of corrosion.

In the following, only selective oxidations will be examined, that is to say when the chemical phenomena has to be controlled (nature of the product, productivity and selectivity).

It is paradoxical to note that oxygen (atmospheric or pure) has been utilized rather recently as oxidizing agent.

During a long time, oxidation processes were based on electrochemistry and some reactions of carbochemistry and used reactants such as chlorine and nitric acid (SCHIRMANN [1]).

Now, atmospheric oxygen is by far the least expensive oxidizing agent and consequently the most utilized. However the operating conditions correspond often to low productivity and high installed capacities.

Reaction mechanisms

Two types of mechanisms, homolytic and heterolytic, are involved in the reactions of oxidation of organic products.

For more details see the review published by G. FRANZ and R.A. SHELDON [2].

In the following only gas-liquid oxidation with air corresponding to the homolytic mechanism will be examined. In table 1, the main industrial processes of this type are given. They involve generally as intermediary state, the production of hydroperoxides or peracids. Globally this represents a very large chemical industrial production.

Chemical reaction engineering

Briefly, it can be recalled that this term includes all the aspects which have to be taken into account for choosing the reactor type and the operating conditions (table 2).

Table 2

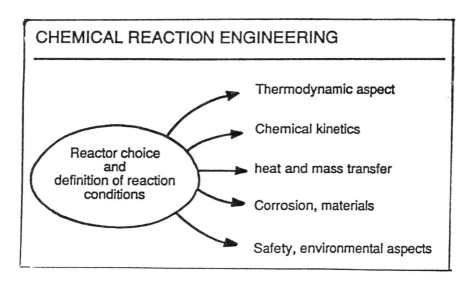

Table 1

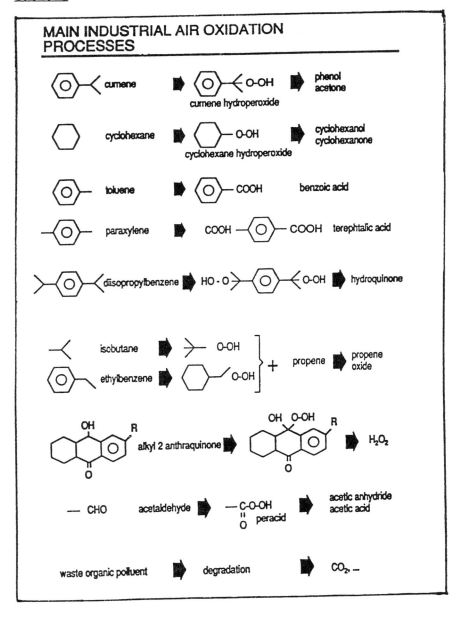

There is a very large interaction between these different parameters, and the choice of the reactor results from a compromise between various constraints. Particularly, very important safety problems are related to the ignition (or autoignition) of the mixtures of air and organic products and to the risks due to the decomposition of hydroperoxides which are highly unstable products. This leads to the necessity of controlling, among other parameters, the oxygen concentration in the reactant mixture and the reaction temperature. This strict control brings about consequences on the mass transfer problems.

To solve this various contraints, several reactions technologies are available. As example, we have schematized bubble column reactors on figure n° 1.

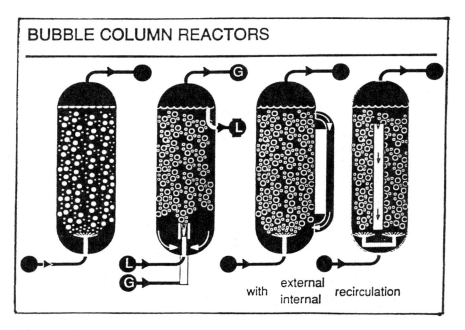

BUBBLE COLUMN REACTORS

with external / internal recirculation

Figure 1

In the following, we will pay attention to only one of these aspects, i.e. the modelisation of the chemical transformation taking account :

- the kinetics of reaction
- the hydrodynamic and mass transfer problems.

Reaction regime

The regime can be :

- either kinetic (or chemical) in which case the kinetics of the chemical reaction controls the result of the chemical transformation

- or diffusional (mass transfer regime) in which case the mass transfer rate is the controlling parameter. The experimental regime is often intermediary between then two extreme cases.

A simple method to get some knowledge about the nature of the reaction regime is to compare the characteristic times of the system :

i) the reaction time t_r which is related to the reaction rate (for example the time required to get a conversion of 80 % in a batch reactor).

ii) the mean residence time of the liquid $\theta = \dfrac{V_L}{Q_L}$

V_L = liquid volume
Q_L = liquid flow rate

iii) the characteristic time of the mass transfer

$$t_{MT} = \frac{1}{k_L a}$$

which will be defined more precisely below.

Lower is T_{MT}, better is the mass transfer.

In figure n° 2, the principles of this method are given. In the kinetic regime, a simple hydrodynamic model is sufficient (eg CSTR) when the reaction time is much higher than the mean residence time.

In the other case, a detailed analysis of the hydrodynamics is required.

Real cases are very often not so contrasted and characteristic times are comparable (of the same order of magnitude). It is then necessary to carry out a precise quantitative comparison of the rate of mass transfer with the rate of the reaction. The rate of the mass transfer can be written :

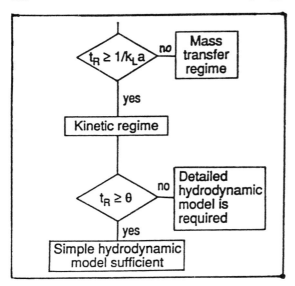

<u>Figure 2</u>

$R = K_La (C^*_{O2} - C_{O2})$
(moles/surface.time)

This relation includes a second important factor which is $C^*_{O2} - C_{O2}$ (see fig. 3 for the details of the basic equations of the mass transfer).

As example, for the methylethylketone oxidation, HOBBS et al. [3] have shown that according to the operating conditions the reaction regime can be modified between the bottom and the top of a reactor (bubble column type):

— lower part :
  $C_{O2} \approx C^*_{O2} \rightarrow C^*_{O2} - C_{O2}$ small

  $\rightarrow$ kinetic regime

— higher part :

  $C_{O2} \leq C^*_{O2} \rightarrow C^*_{O2} - C_{O2}$ high

  $\rightarrow$ diffusional regime

  (see figure n° 4 a)

BASIC MASS TRANSFER EQUATION

$$N = K_L (Co_2^* - Co_2)$$

(modes / Surface.time)

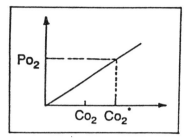

$$Co_2 \quad Po_2$$

$$Co_2^* = \frac{Po_2}{He}$$

K_L Overall volumetric mass transfer coefficient

He   HENRY coefficient measures the solubility of oxygen in the liquid

for example O_2 in hydrocarbons 0.005 to 0.015 mole/l.

Rate of production $R = K_L\, a\, (Co_2^* - Co_2)$
(e. g. moles / volume.time)

a specific interfacial area

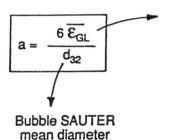

$$a = \frac{6\,\overline{\mathcal{E}_{GL}}}{d_{32}}$$

Bubble SAUTER mean diameter

Gas hold-up in the reactor

$$\overline{\mathcal{E}_{GL}} = \frac{V_G}{V_G + V_L}$$

V_G, V_L volumes of gas and liquid

RF "ROUSING factor"

$$RF = \frac{1}{1 - \overline{\mathcal{E}_{GL}}}$$

Figure 3

WHICH REACTION REGIME ?

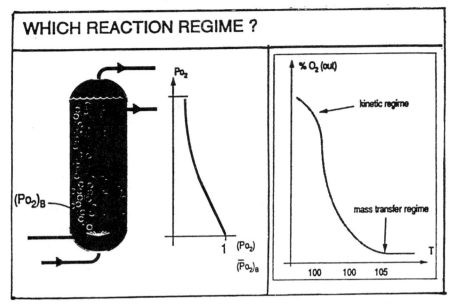

Figure 4 (a)   Figure 4 (b)

The reaction regime can change as a function of the temperature

—   low temperature
     slow chemical reaction
     → kinetic regime

-   high temperature
     fast chemical reaction
     → diffusional regime.

     (see figure n° 4 b)

Volumic conductances of gas/liquid transfer $k_L a$

It is the main parameter for the quantification of the intensity of a G/L transfer. Its determination is delicate since it is related to very different factors (see table 3).

k_L	$a = \dfrac{6 \cdot \varepsilon_{GL}}{d_{32}}$
– physical properties of the liquid (viscosity ..) – gas diffusivity in the liquid – turbulent energy dissipation	– physical properties of the phases – breakage/coalescence behaviour – turbulent energy dissipation

The influence of these factors can be illustrated for two of them :

- the energy dissipation and extrapolation problems

- the characteristics of the breakage/coalescence properties of the utilised gas/liquid system.

a) Influence of the energy dissipation.

The energy dissipation is characterized by the parameter

$$\varepsilon = \frac{p}{M}$$

p = dissipated power
M = gas/liquid dispersion mass

In many cases, the energy dissipation in a reactor is very heterogeneous. In figure n° 5 the local variations of energy dissipation are represented in the highter part of a reactor stirred by a turbine (from BARTHOLE and al [4]).

$\varepsilon^* = \dfrac{\varepsilon}{\bar{\varepsilon}}$, where $\bar{\varepsilon}$ represents the average value of the energy dissipated in the tank, can vary between 0,21 to 2,4, i.e. by a factor of 10.

The difficulty of defining a global k_La appears since the parameters is related to the local dissipation of energy.

This fact has another consequence at the extrapolation level when the local dissipation can be modified as a function of increase of the reactor sizes which will introduce another error sources.

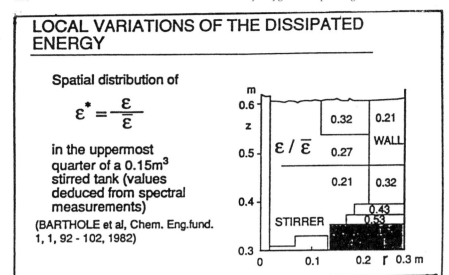

Figure 5

Two tendencies appear now :

* Definition of zones corresponding to different k_La linked with recirculation flows instead of calculating an average value of k_La for the whole reactor.

* In the long term, development of computing flow hydrodynamic methods. In the meantime, the hydrodynamics (gas holdup, phases velocity) of gas liquid reactors can be simulated for simple axisymmetric geometries (PETERSEN [5]).

b)   Effect of breakage/coalescence properties.

The interfacial area is strongly dependent upon the breakage/coalescence properties of the G/L system. All other parameters being equivalent, the mean diameter of bubbles can vary from 1 to 10 according to the nature of the reactor. The physicochemical properties of the phases (density, viscosity, surface tension) are not the only paramaters to intervene. The mechanisms are complicated and a lot of work would be necessary to clarify all these points.

All the same, it is possible to simulate these behaviour in a cold pilot model.

Laboratory studies in a bubble column has allowed the characterisation of various G/L system by a mean bubble diameter and their simulation by an aqueous solution of an alcohol (butanol, ethanol) adjusting the alcohol concentration, (see figure n° 6). It is then possible to proceed to optimization and extrapolation studies in a pilot model using theses aqueous solutions.

This type of study is often essential, not only to account for the breakage coalescence properties but also for the specific geometries of reactors.

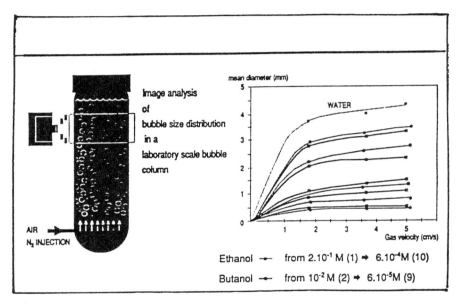

Ethanol $\rightarrow$ from 2.10^{-1} M (1) $\rightarrow$ 6.10^{-4} M (10)

Butanol $\rightarrow$ from 10^{-2} M (2) $\rightarrow$ 6.10^{-5} M (9)

Figure 6

However some general data are available in the literature. They allow the comparison of the performances of different set-up. For example, we can mention the KEITEL and ONKEN diagram, revised by VILLERMAUX [7] which represents $k_L a$ in a dimensionless form according to the dimensionless form proposed par ZLOKARNIK. (See figure n° 7).

Overall modelisation

The final objective remains the comprehensive simulation of the chemical and mass transfer kinetic data.

One of the first examples of such studies, published in the literature, is the work of CHANESE et al.[8] in the case of cumene oxidation.

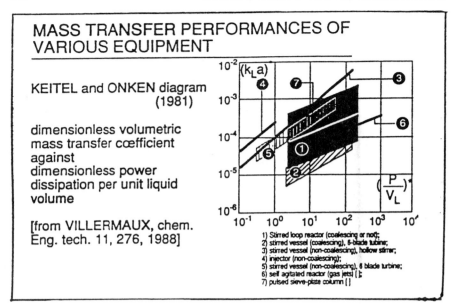

MASS TRANSFER PERFORMANCES OF VARIOUS EQUIPMENT

KEITEL and ONKEN diagram (1981)

dimensionless volumetric mass transfer coefficient against dimensionless power dissipation per unit liquid volume

[from VILLERMAUX, chem. Eng. tech. 11, 276, 1988]

1) Stirred loop reactor (coalescing or not);
2) stirred vessel (coalescing), 6-blade turbine;
3) stirred vessel (non-coalescing), hollow stirrer;
4) injector (non-coalescing);
5) stirred vessel (non-coalescing), 6 blade turbine;
6) self agitated reactor (gas jets) [];
7) pulsed sieve-plate column []

Figure 7

This author solved the equations of a model including :

-   a complex kinetic mechanism (proposed by HATTORI et al. [9]).

-   the hydrodynamics of a bubble column with a back flow cell model (see figure 8).

Now the recent computing progress permits us to solve very complicated equations systems and consequently more and more ambitious models can be proposed.

To illustrate the last point, two very recent papers can be mentioned (see figure n° 9) :

-   the work of POHORECKI et al [10] related to the modeling of a cyclohexane oxidation process.
-   the study of ANDRIGO et al [11] related to the modeling of a cumene oxidation process.

In both cases, the results of the industrial simulations are satisfactory.

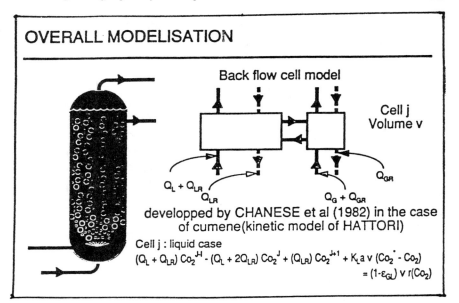

OVERALL MODELISATION

Back flow cell model

Cell j
Volume v

developped by CHANESE et al (1982) in the case
of cumene(kinetic model of HATTORI)

Cell j : liquid case

$$(Q_L + Q_{LR}) Co_2^{J-1} - (Q_L + 2Q_{LR}) Co_2^{J} + (Q_{LR}) Co_2^{J+1} + K_L a \, v \, (Co_2^{*} - Co_2)$$
$$= (1 - \varepsilon_{GL}) \, v \, r(Co_2)$$

Figure 8

Conclusion

The oxidation reactions with air illustrate perfectly
the different aspects of chemical engineering and their
interactions.

The computing progress will allow us to get further in
the knowledge of hydrodynamics of reactions using
numerical hydrodynamic simulations and in the
realisation of global models taking into account
complicated kinetical schemes.

MODELISATION OF THE CYCLOHEXANE OXIDATION PROCESS

CYCLOPOL process [POHORECKI and al., Chem. Eng. Sci 47, 9-11, 2559, 1992]

Development of a model including both reaction kinetics
(12 basic equations) and mass transfer

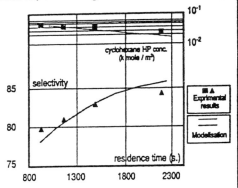

The kinetic parameters have been determined from literature data and laboratory experiment results

Industrial simulation results
$(V_{TOT} = 110 \ m^3, 5 \ CSTR \ in \ series)$

MODELISATION OF THE CUMENE OXIDATION PROCESS

[ANDRIGO and al., chem. Eng. Sci, 47, 9 - 11, 2511, 1992]

Development of a kinetic model based on 8 classical reactions steps + 10 side reactions steps

The kinetic parameters (22) have been determined from results of a laboratory study in a CSTR

		1	2	3	4	5	6	7	8
CHP selectivity (% mol.)	EXP	93,9	91,4	90,5	90,5	88,5	87,1	87,6	85,6
	CAL	93,7	91,2	90,7	90	89	88,2	87,1	85,9

INDUSTRIAL SIMULATION RESULTS (ENICHEM)
$(V_{TOT} = 400 \ m^3$, 8 bubble columns with internal recirculation in series)

Figure 9

References

[1]   J.P. Schirmann
      Oxydations ménagées (Technique de l'ingénieur,
      procédés unitaires)

[2]   G. Franz and R.A. Sheldon
      'Oxidation', ULLMANNN'S encyclopedia of indus.
      chem.

[3]   C.C. Hobbs, E.H. Drew, H.A. Vant'hof, F.G. Mesich
      and M.J. Onore
      IEC. prod. res.dev., 11, 2, 220, 1972

[4]   J.P. Barthole, J. Maisonneuve, J.N. Gence, R.
      David, J. Mathieu and J. Villermaux
      Chem. Eng. Fund. 1, 1, 17, 1982

[5]   K. Petersen
      Thèse, Université de Lyon I (1992)

[6]   O. Coccaud, C. Mathieu, R.V. Chaudari et J.
      Breysse
      Rhône-Poulenc Internal Report

[7]   J. Villermaux
      Chem. Eng. Tech., 11, 276, 1988

[8]   A. Chanese
      Chem. Eng. Comun., 17, 261, 1982

[9]   K. Hattori, Y. Tanaka, H. Suzuki, T. Ikawa and H.
      Kubota
      J. Chem. Eng. Jap. 3, 1, 72, 1970

[10] R. Pohorecki, J. Baldyga, W. Moniuk,
      A.Krzysztoforski and Z. Wojcik
      Chem. Eng. Sci., 47, 9-11, 2559, 1992

[11] P. Andrigo, A. Caimi, P. Cavalieri D'oro, A. Fait,
      L. Roberti, M. Tampieri and V. Tartari
      Chem. Eng. Sci, 47, 9-11, 2511, 1992

Electropox: BP's Novel Oxidation Technology

T. J. Mazanec, T. L. Cable, and J. G. Frye, Jr.
BP RESEARCH AND ENVIRONMENTAL SCIENCE CENTER, 4440
WARRENSVILLE CENTER ROAD, CLEVELAND, OHIO 44128, USA

1 INTRODUCTION

Oxidation processes account for the preparation of a great variety of chemicals and fuels. Engineering of the admixing of oxygen and hydrocarbon in these processes has generated a number of different reactors and concepts that allow selective reactions to be conducted that would otherwise remain unselective. A design that has only recently received significant attention in this regard is the membrane reactor concept in which oxygen separation from air and oxidation catalysis are conducted simultaneously.

A particularly challenging oxidation that we have studied in membrane reactors is the partial oxidation of methane. Methane, the chief constituent of natural gas, is a very robust hydrocarbon. Products of methane reactions are generally more reactive than methane, limiting schemes for direct methane conversion to useful materials.[1-2] With this in mind, attempts to upgrade methane have begun to focus on the relative rates of oxidation of methane compared to its products[3-4] and attention has been directed toward the fundamental kinetic limitations of conventional processes. Gas phase molecular oxygen often has been cited as the species responsible for the destructive oxidation of the desired products. If this is the case a membrane reactor has the potential to overcome the kinetic limitation of the methane oxidation process by keeping free molecular oxygen separated from the products.

Solid oxide fuel cells can be considered to be simple membrane reactors wherein oxygen is conducted from air to fuel in a controlled fashion while the bulk of the air and fuel are separated by the electrolyte. A number of studies of the methane oxidation reaction have been reported using solid oxide fuel cells.[5-18] Results of these studies demonstrate that fuel cells can act as oxygen membranes to produce useful products from methane.

The present contribution considers the important question of whether the promise of enhanced selectivity can be achieved in methane oxidation using oxygen membrane reactors. Results are presented from studies on fuel cells as the oxygen separating device and from studies using advanced membrane concepts to separate oxygen from air.[19-22] An exciting new application of these devices has been discovered to be the partial oxidation of methane to synthesis gas in a process referred to as Electropox.[23]

2 EXPERIMENTAL

Reactor Operation

External circuit reactions were carried out in either a tubular or disc reactor. The tubular reactor is shown in Figure 1. The reactor consists of a yttria stabilized zirconia tube coated over a central section 100 mm in length inside and out with the cathode and anode material, respectively. Electrical contact is made to the electrodes by stripes of Ag (or Pt) painted along the length of the YSZ tube. The tube fits into a quartz sleeve to form a shell and tube heat exchanger arrangement. The disc reactor is shown in Figure 2. A disc is fitted between two supporting ceramic tubes and contacted on either side by a flanged SS tube that acts as both gas feed tube and electrical contact. Internally shorted, dual phase cell discs are fitted into the same reactor, except the flanged SS tube is replaced by either a stubbed SS or mullite tube. The reactors are heated in a split furnace.

Details of reactor operation and cell preparation can be found elsewhere.[23]

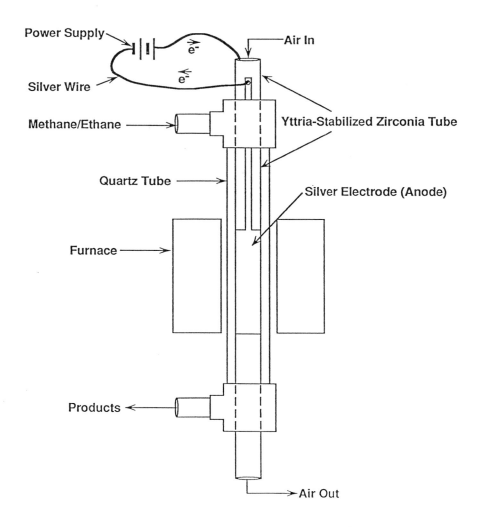

<u>Figure 1</u>   Tubular Electrocatalytic Reactor

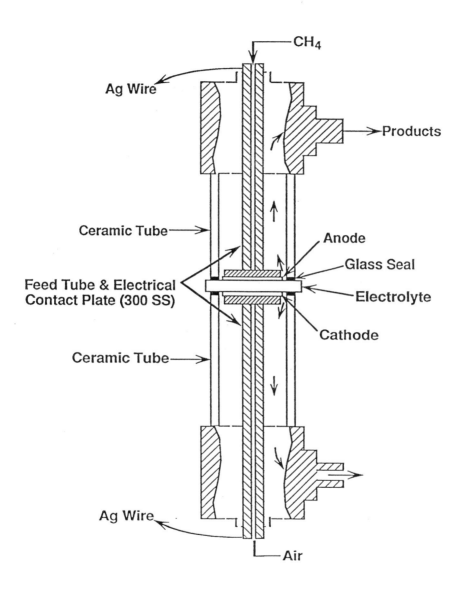

Figure 2   Electrocatalytic Disc Reactor

3   RESULTS AND DISCUSSION

Methane oxidation reactions are strongly temperature dependent. Table 1 summarizes the effect of temperature on the conversion of methane in a short circuited electrochemical cell reactor. As expected, the current, which is a measure of oxygen transport, increases with temperature due to the increased conductivity of YSZ at higher temperatures. This means that the oxygen/methane ratio is also increasing with temperature in these experiments. The selectivity to C2's increases up to 750°C and then appears to reach a plateau. The fraction of C2 product that appears as ethylene likewise increases with temperature. None of these observations is inconsistent with other studies of oxidative methane coupling.

The effect of residence time on the methane oxidation reaction in the tubular electrocatalytic reactor was investigated by subjecting a Ag anode cell to different flow rates of methane as shown in Table 2. In these short circuit experiments the current is held relatively constant as the gas flow rate is changed, resulting in a change in the CH_4/O_2 ratio. As the CH_4/O_2 ratio is decreased the conversion increases and selectivity to C2's declines, similar to what has been observed in heterogeneously catalyzed methane coupling experiments.

Table 1   Effect of Temperature with AgPbMn/YSZ/Ag Cell

Temp.	Current	CH_4 Conv	Molar Selectivity, (%)			
(°C)	(mA)	(%)	C_2H_6	C_2H_4	C_3H_6	CO_2
700	58	1.0	48.9	—	—	51.1
725	79	1.4	42.7	11.8	—	45.5
750	99	1.8	41.7	16.6	—	41.7
775	129	2.5	37.5	21.8	—	40.7
800	177	3.3	32.0	26.4	—	41.6
825	226	4.3	24.9	28.9	4.1	42.1

CH_4 Feed Rate = 20 cc/min

Table 2 Effect of Residence Time at Constant Current

Res. Time	Current	CH_4/O_2 Ratio	CH_4 Conv	Molar Selectivity, (%)			
(sec.)	(mA)		(%)	C_2H_6	C_2H_4	C_3H_6	CO_2
2.8	125	47.3	3.2	29.7	30.8	5.5	34.0
7.1	126	18.4	6.4	15.6	31.2	5.7	47.5
12.7	130	10.2	10.6	9.2	25.2	6.5	59.1

Temperature = 800°C, Ag/YSZ/Ag Cell

Variation of residence time at constant CH_4/O_2 ratio was achieved by applying an adjustable voltage and controlling the methane feed rate with mass flow controllers. Results of these experiments, summarized in Table 3, demonstrate that the methane conversion and selectivity to C2's remain nearly constant as the residence time is changed under constant CH_4/O_2 ratio conditions. The ratio of ethylene to ethane increases with residence time which is attributed to increased thermal pyrolysis. These experiments were conducted with a cell containing a Bi-doped Ag anode.

Table 3 Effect of Residence Time with Constant CH_4/O_2 Ratio

Res. Time	Current	CH_4/O_2 Ratio	CH_4 Conv	Molar Selectivity, (%)			
(sec.)	(mA)		(%)	C_2H_6	C_2H_4	C_3H_6	CO_2
2.9	343	16.7	7.3	23.8	29.5	4.2	42.1
4.9	206	16.4	7.2	19.6	33.9	5.5	41.0
7.9	128	16.7	6.9	17.6	32.2	5.9	44.3

Temperature = 800°C, AgBi/YSZ/Ag Cell

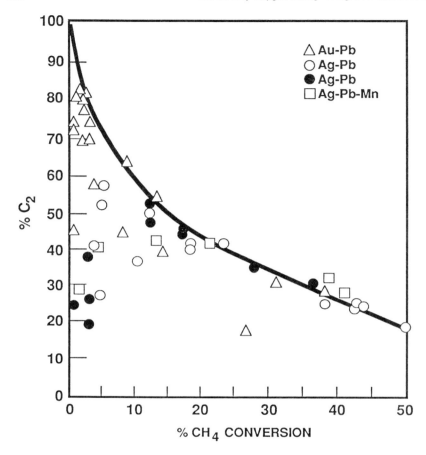

<u>Figure 3</u>   Methane Coupling Selectivity and Conversion

$$A \;\overset{k_1}{\dashrightarrow}\; B \;\overset{k_2}{\dashrightarrow}\; C$$

$$CH_4 \dashrightarrow C_2 \dashrightarrow CO_2$$
$$\downarrow$$
$$CO_x$$

<u>Figure 4</u>   Consecutive Reaction Mechanistic Scheme

The rate constants, k_1 and k_2 in the Figure 4, determine the upper limit of C2 yield. (Since ethane can convert to ethylene readily by non-oxidative pyrolysis and ethylene is more stable than ethane the rate of ethylene activation is considered to be the limiting feature.) This is the best of all possible worlds as far as catalysis is concerned, since incorporation of a direct $CH_4 \rightarrow CO_2$ step can only decrease C2 yield.

Lunsford[24,25] and others have determined that homolytic C-H bond rupture is the rate limiting step in CH_4 activation. Since there are no other available chemical means to activate or complex methane except the very weak and inconsequential van der Waals forces, the active site on the catalyst that attacks this C-H bond can be assumed to be capable of attacking any C-H bond. Thus in the rate equations for CH_4 and C_2H_4 activation the same concentration of oxidation sites, [Ox], can be used:

$$\text{Rate } CH_4 = - k_1 [CH_4] [Ox] \tag{1}$$

$$\text{Rate } C_2H_4 = - k_2 [C_2H_4] [Ox] \tag{2}$$

This greatly simplifies analysis of the system since the two rates differ only in the fundamental rate constant and the concentration of the reactant. Computer simulation of the "ABC" mechanism allows one to determine the k_2/k_1 ratio needed to achieve the desired yield of C_2H_4. Figure 5 shows how the selectivity to B varies as a function of conversion of A for various k_2/k_1 ratios. (Since oxidative activation of ethane to give an ethyl radical results in an ethylene product, it only serves to consume oxidant and can be ignored. If direct ethane conversion to CO_2 occurs this will reduce C2 yield.) The results in Figure 3 are fit well by a k_2/k_1 ratio of about 6.

Experimental test of this mechanism was conducted by admitting ethylene/methane mixtures into the tubular reactor. The results, summarized in Table 5, demonstrate that ethylene oxidation competes readily with methane oxidation under the experimental conditions of the electrocatalytic cell. The ratios of k_2/k_1 calculated for these experiments are 4.0 and 4.6. This is in reasonable agreement with the ratio derived from the methane coupling experiments. Thus, the ABC mechanism can be applied successfully to systems of this type.

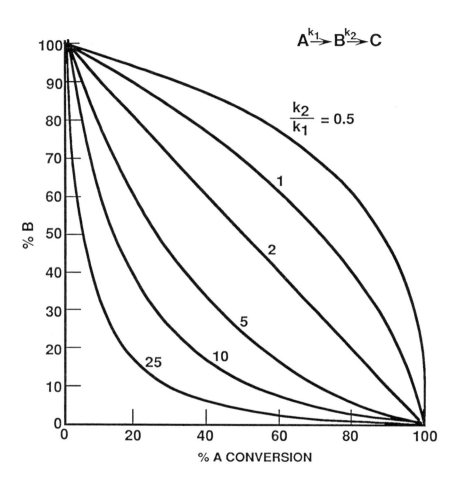

Figure 5   Selectivity/Conversion Relation as a Function of k_2/k_1

Table 5 Ethylene Cofeed Experiment

Feed (Mole %)		Current	Effluent (mole %)		K_2/K_1
CH_4	$C_2=$	(mA)	CH_4	Ethylene	
100	—	885	82.4	3.9	–
91.4	8.7	620	86.4	6.8	4.0
82.7	17.3	490	79.5	14.2	4.6

Temperature = 800°C, AgPb/YSZ/Ag Cell
CH_4 Feed Rate = 20 cc/min, Applied Voltage = 2.0 V

At higher temperatures, using tubular cells with Pt anodes, the conversion of methane becomes more selective with CO as the major product. As summarized in Table 6 methane conversions can approach 100% with CO selectivities up to 97% at 1100°C. The electrocatalytic process used for the partial oxidation of methane to synthesis gas we call Electropox.

The limitation of the tubular reactor is that the current densities (oxygen fluxes) are very low. This is due in part to the thick tube wall (1-1.5 mm) and in part to the poor adherence of the electrodes to the

Table 6 Effect of Temperature on Methane Conversion

Temp °C	CH_4 Rate cc/min	Applied Potential Volts	Current mA	CH_4 Conv %	Product Selectivity		
					C2's	CO	CO_2
800	20.6	0.0	107	3.1	22.6	56.6	20.8
800	8.0	0.0	227	23.5	–	95.9	4.1
1000	8.0	0.0	340	40.4	–	99.9	0.1
1100	8.0	0.0	330	39.2	–	99.9	0.1
1100	8.0	2.0	994	99.9	–	96.8	3.2

Cell: CH_4//Pt-YSZ/YSZ/Pt//Air

Table 7 Current Density, Syngas Yields in Short Circuit
Cells.

Cell	Feed	Current Density mA/cm^2	Resistance Ω-cm^2	CO Yield
Pt/YSZ-4/Pt (tube)	CH$_4$	15	50	99
Pt/YSZ-4/LSM	CH$_4$	347	2.65	-
PtBi/YSZ-4/LSM	CH$_4$	525	1.34	56
PtBi/YSZ-8/LSM-LSM	CH$_4$	552	1.25	61
PtBi-cermet/YSZ-8/LSM	CH$_4$	705	0.8	-
Ni-cermet/YSZ-8/LSM-Pt	CH$_4$	930	-	83
	H$_2$	1245	0.45	-

tube, resulting in very large contact resistances. By
utilizing an externally shorted disc reactor the
interfacial resistance could be greatly reduced as shown
in Table 7. Current densities of greater than 1 A/cm^2
were obtained. Increasing the oxygen flux greatly
reduces the size of the reactor required to convert a
given quantity of natural gas.

Another way of improving the oxygen flux is to
introduce an internal short circuit. Since for most
natural gas upgrading schemes electricity is not a
valuable product due to the remote location of the
plant, the external circuit can be discarded. A short
circuit is obtained by introducing into the ionically
conducting electrolyte a second phase to conduct
electrons. This is demonstrated schematically in Figure
6. Oxide ions can migrate from air to natural gas via
the ionically conducting phase and electrons can return
via the electronically conducting phase.

Results with these dual phase membranes have been
impressive. Table 8 summarizes some of the oxygen flux
results obtained with dual phase membranes separating
hydrogen and air at 1100°C. Ease of fabrication and
high oxygen flux make these ideal candidates to form
into membranes of all shapes for conducting the
Electropox process.

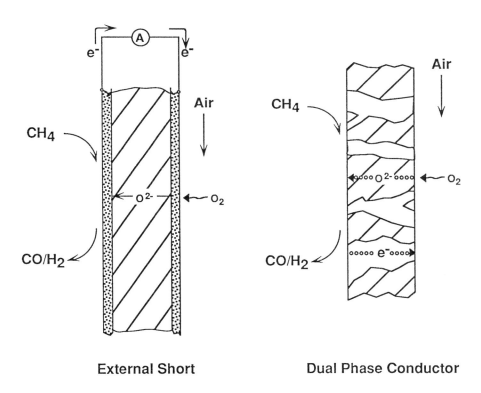

External Short   **Dual Phase Conductor**

<u>Figure 6</u> Externally Short Circuited Cells and Dual
Phase Membranes

The Role of Oxygen in Improving Chemical Processes

Table 8 Oxygen Flux Results for Dual Phase Membrane Discs at 1100C

Membrane Electronic/Ionic Phase	Feed	Current Density mA/cm^2	Oxygen Flux $cc/min-cm^2$	Thickness mm
$(Pd)_{.5}/(YSZ)_{.5}$	H_2	555	2.1	0.8
	CH_4	530	2.0	0.8
$(Pt)_{.5}/(YSZ)_{.5}$	H_2	467	1.8	0.8
$(B\text{-}MgLaCrO_x)_{.5}/(YSZ)_{.5}$	H_2	114	0.4	0.8
$(In_{90}Pr_{10})_{.4}/(YSZ)_{.6}$	H_2	275	1.1	0.8
$(In_{90}Pr_{10})_{.5}/(YSZ)_{.5}$	H_2	601	2.3	0.8
	H_2	1458	5.5	0.3
	H_2	1611	6.1	0.25
$(In_{95}Pr_{2.5}Zr_{2.5})_{.5}/(YSZ)_{.5}$	H_2	2083	7.8	0.3

4 CONCLUSIONS

1. Electrocatalytic methane upgrading can achieve modest yields of ethane and ethylene.

2. Methane coupling yields are limited by the subsequent oxidation of the C2 products.

3. The ABC mechanism permits one to assess the yields of ethylene attainable from a methane coupling process.

4. High temperature methane upgrading to synthesis gas via an electrocatalytic process (Electropox) can result in high yields of CO.

5. High oxygen fluxes are able to be obtained in externally short circuited cells.

6. Dual phase, internally short circuited membranes can be fabricated that give high oxygen flux.

ACKNOWLEDGEMENT

We thank the British Petroleum Company, PLC for permission to publish this manuscript.

References

1. J.A.Labinger and K.C.Ott, <u>J. Phys Chem,</u>, 1987, <u>91</u>, 2682.
2. J.A.Labinger, <u>Cat. Lett.</u>, 1988, <u>1</u>, 371.
3. R.Burch and S.C.Tsang, <u>Appl. Cat.</u>, 1990, <u>65</u>, 259.
4. P.R.Pereira, V.De Gouveia, F.Rosa, <u>Preprints Petr. Div. ACS</u>, 1992, <u>37</u>, 200.
5. R.A.Goffe, D.M.Mason, <u>J. Appl. Electrochem.</u>, 1981, <u>11</u>, 447.
6. B.G.Ong, C.C.Chiang, D.M.Mason, <u>Sol. State Ionics</u>, 1981, <u>3/4</u>, 447.
7. B.C.Nguyen, T.A.Lin, D.M.Mason, <u>J. Electrochem. Soc.</u>, 1986, <u>133</u>, 1807.
8. K.Otsuka, S.Yokoyama, A.Morikawa, <u>Bull. Chem. Soc. Jpn.</u>, 1984, <u>57</u>, 3286.
9. K.Otsuka, S.Yokoyama, A.Morikawa, <u>Chem. Lett. (Japan)</u>, 1985, 319.
10. K.Otsuka, A.Morikawa, Japan Pat. 61-30688, 12 Feb 1986.
11. K.Otsuka, K.Suga, I.Yamanaka, <u>Catal. Today</u>, 1990, <u>6</u>, 587.
12. M.Stoukides, C.G.Vayenas, <u>J. Catal.</u>, 1980, <u>64</u>, 18.
13. S.Seimanides, M.Stoukides, <u>J. Electrochem. Soc.</u>, 1986, <u>133</u>, 1535.
14. D.Eng, M.Stoukides, <u>Catal. Lett.</u>, 1991, <u>9</u>, 47.
15. N.Kiratzis, M.Stoukides, <u>J. Electrochem. Soc.</u>, 1987, <u>134</u>, 1925.
16. H.Nagamoto, K.Hayashi, H.Inoue, <u>J. Catal.</u>, 1990, <u>126</u>, 671.
17. D.J.Kuchynko, R.L.Cook, A.F.Sammells, <u>J. Electrochem. Soc.</u>, 1991, <u>138</u>, 1284.
18. N.U.Pujare, R.L.Cook, A.F.Sammells, US 4,997,725, 5 Mar 1991, assigned to Gas Research Institute.
19. T.J.Mazanec, T.L.Cable, US 4,802,958, 7 Feb 1989, assigned to the Standard Oil Co.
20. T.J.Mazanec, T.L.Cable, J.G.Frye, Jr., US 4,793,904, 27 Dec 1988, assigned to the Standard Oil Co.
21. T.J.Mazanec, T.L.Cable, US 4,933,054, 12 Jun 1990, assigned to the Standard Oil Co.
22. T.L.Cable, T.J.Mazanec, J.G.Frye, Jr., Eur. Pat. Appl. 0399833, 28 Nov 1990, assigned to The Standard Oil Co.
23. T.J.Mazanec, T.L.Cable, J.G.Frye, Jr., <u>Solid State Ionics</u>, 1992, in press.
24. D.J.Driscoll, J.H.Lunsford, <u>J. Phys. Chem.</u>, 1985, <u>89</u>, 4415.
25. J.H.Lunsford, <u>Catal. Today</u>, 1990, <u>6</u>, 235; and references therein.

New Approaches to Large-scale Selective Oxidation Processes

R. Ramachandran
THE BOC GROUP TECHNICAL CENTER, MURRAY HILL, NEW JERSEY
07974, USA

1   INTRODUCTION

In the trillion dollar Chemical Industry[1], petrochemical production accounts for about half the sales. In the petrochemical industry, selective oxidation is one of the important processes and it accounts for about 20%[2] of the total output of the major organic chemicals. Selective oxidation processes differ from energy conversion processes in that it converts the feed hydrocarbons to valuable products in presence of a catalyst while minimizing the formation of CO and CO_2.

Selective oxidation processes are carried out in presence of a catalyst to maximize the formation of the desired product. These reactions are carried out using both homogeneous and heterogeneous catalysis.

Homogeneous catalysis include: selective oxidation with hydroperoxides and hydrogen peroxide; liquid phase oxidations; asymmetric oxidation; direct olefin amination; and oxycyanation.

Heterogeneous catalytic oxidation processes are widely used to produce a variety of chemicals and include: allylic oxidation to form aldehydes, nitriles, anhydrides and acids; aromatic oxidation to form acids and anhydrides; epoxidation of olefins to oxides; methanol oxidation to formaldehyde; and paraffin oxidation to anhydrides and nitriles; oxidative dehydrogenation of butene to butadiene.

2   SELECTIVE OXIDATION PROCESSES

Several important organic chemicals are produced by heterogeneous selective oxidation catalysis. Many of these processes have been in commercial operation and improvements in processes and catalysts are always in progress to further improve existing processes and to develop new processes. Generally, these processes are based on metallic oxides, either supported or unsupported, as catalysts. Also, these catalysts contain a variety of promoters to increase their activity, selectivity, strength and life. Typical processes and the catalysts employed are shown in Table 1.

3 CATALYTIC CHEMISTRY

In some of the above reactions, an oxygen-containing gas is introduced directly as a feed into the process while in the others oxygen is incorporated into the reaction through a carrier, such as peroxides in the production of propylene oxide. In most of the cases the catalyst is maintained in an oxidized form during the reaction.

The catalytic oxidation reactions can be classified as electrophilic oxidation, in which the activation of oxygen is the first step, and this followed by nucleophillic oxidation, which proceeds through the activation of the hydrocarbon molecule[3]. An example of a reaction in which activation of oxygen occurs first is oxidation of olefins to oxides (ethene to ethylene oxide) while in the oxidation of olefins to anhydrides, nitriles etc. activation of the hydrocarbon is the first step and these steps are believed to be the rate controlling steps.

Generally it is believed that the catalytic oxidation reactions are surface processes in which the reactions take place between the gaseous hydrocarbon feed and the active catalytic surface oxygen to form the desired product. Following this, the reduced catalyst surface gets re-oxidized by the gaseous oxygen to the oxidized state. These processes are known as the "Redox mechanism"[4] and were first identified by Mars and van Krevelan[4]. A redox process is shown in Figure 1. In this, the reactant hydrocarbon is adsorbed on M_1^{n+} site to form the chemisorbed species and reacts with the lattice oxygen associated with the M_1^{n+} site to produce the partially oxidized product. During this phase, a lattice oxygen from a neighboring M_2^{m+} site moves to a M_1^{n+} site to replenish the lost oxygen and electrons move to M_2^{m+} from M_1^{n+}. Subsequently, molecular oxygen reacts with M_2^{m+} to restore its oxidation state.

Various studies have been performed to confirm the above mechanism. One of them is by Keulks et al.[5] in the oxidation of propene. While the reactor was operating at a steady state at a temperature of around $450^{\circ}C$, they switched the oxidant feed from oxygen to $^{18}O_2$ and monitored the products using a mass spectrometer and found that 98% of the products,

Table 1

Product	Raw Materials	Catalyst	Reaction conditions
Acetaldehyde	Ethene, O_2	$CuCl_2/PdCl_2$ (Homogeneous)	$120^{\circ}C$; 3 bar P
Acrylic Acid	Propene, O_2	Oxides of Mo or Sb (Supported)	$250^{\circ}-400^{\circ}C$; 1-3 bar P
Acrylonitrile	Propene or Propane, O_2, NH_3	Oxides of Mo or Sb (Supported)	$400-500^{\circ}C$; 1-2 bar P
Benzoic Acid	Toluene, O_2	Co salt (Homogeneous)	$110-120^{\circ}C$; 3 bar P
Ethylene Oxide	Ethene, O_2	Ag (Supported)	$230-290^{\circ}C$; 10-25 bar P
Formaldehyde	Methanol, O_2	Oxides of Mo and Fe (Supported) or Ferric Molybdate	$300-400^{\circ}C$; 0.5-1 bar P
Maleic Anhydride	Butane or Butene or Benzene, O_2	Vanadia (Supported)	$350-500^{\circ}C$; 1-3 bar P
Phthalic Anhydride	o-Xylene, O_2	Vanadia/Titania (Supported)	$350-500^{\circ}C$; 1-3 bar P
Propylene Oxide	Propene, Hydroperoxide	Heavy Metal	
Vinyl Acetate	Ethene, O_2, HOAc	Pd (Supported)	$180-200^{\circ}C$; 1-4 bar P
Vinyl Chloride	Ethene, O_2, HCl	$CuCl_2$ (Supported)	$210-310^{\circ}C$; 1-3 bar P

Note: O_2 source can be either air or O_2.

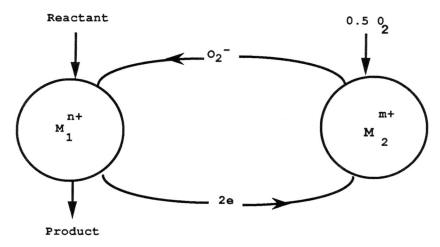

Reactant

Product

FIG. 1. OXIDATION - REDUCTION MECHANISM

[REDOX CYCLE]

acrolein and CO_2, had no $^{18}O_2$ in the product confirming that lattice oxygen is involved in the reaction. Mole percent of reaction products as a function of time during the above experiment is shown in Figure 2.

The Redox mechanism discussed above is explained using reaction kinetics below with propylene oxidation as example:

$$C_3H_6 + \theta_{ox} \xrightarrow{\quad k_{red} \quad} Product + \theta_{red}$$

$$\theta_{red} + O_2 \xrightarrow{\quad k_{ox} \quad} \theta_{ox}$$

$$\theta_{ox} + \theta_{red} = 1$$

where θ_{ox} = fraction of sites which are oxidized
θ_{red} = fraction of sites which are reduced
k_{red} = rate constant for catalyst reduction
k_{ox} = rate constant for catalyst oxidation

Fig. 2 GAS PHASE COMPOSITIONS

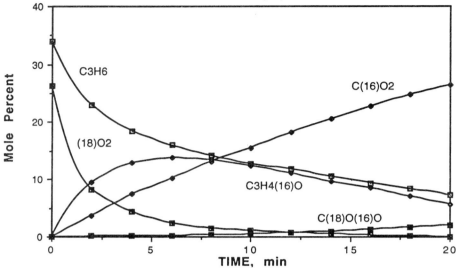

Rate of propylene oxidation =

$$- d(C_3H_6)/dt = k_{red} \cdot p_{C3}^{x} \cdot \theta_{ox}$$

Rate of Oxygen utilization =

$$- d(O_2)/dt = k_{ox} \cdot p_{O2}^{y} \cdot \theta_{red}$$

$$= k_{ox} \cdot p_{O2}^{y} \cdot (1-\theta_{ox})$$

Since the rates of oxidation and reduction are equal under steady state and the number of oxidized states are almost constant when the rate of oxidation is much faster compared to the rate of reduction, one can conclude that when the rate of catalyst re-oxidation is faster than reduction, the rate of propylene oxidation is equal to the rate of catalyst reduction, i.e.

$$- d(C_3H_6)/dt = k_{red} \cdot p_{C3}^{x}$$

Keulks et al.[5-7] also studied the effect of reaction temperature on the rates of propene oxidation on various catalysts and found a distinct change in the mechanism, as shown in Figure 3. The rate of reaction was found to be first order in propene and zero order in oxygen (i.e. the rate of catalyst oxidation is much faster compared to the rate of

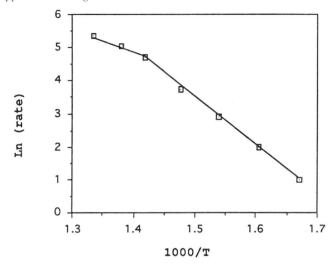

FIG. 3 EFFECT OF TEMPERATURE ON

ACROLEIN FORMATION ON U Sb3 O10

propene oxidation) above a temperature of about $400^{\circ}C$ while at lower temperatures the rate was controlled by the rate of catalyst oxidation.

Hence, it is important, for most of these oxidation reactions, to operate the process where the rate of catalyst oxidation is faster than the rate of reaction so that the life of catalyst will be longer.

4   REACTION MECHANISM

Extensive research has been performed in this area to establish reaction mechanisms for the formation of a variety of chemicals using deuterated hydrocarbons as feed. It has been observed that the product and feed hydrocarbons dissociate to form CO_2 resulting in an sub-optimum performance. At lower feed hydrocarbon conversions, most of the CO and CO_2 are generally formed directly from the feed hydrocarbons while at high conversions, most of the formation is due to the product decomposition. One of the proposed mechanisms for the production of maleic anhydride, for example, is shown in Figure 4.[8,9] Since the maleic anhydride formed can further decompose to CO_2, as conversion of butane increases, selectivity (defined as the ratio of maleic anhydride to total products) to maleic anhydride decreases. At very high butane conversions,

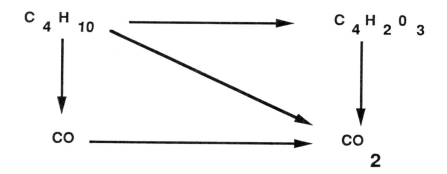

FIG. 4. OVERALL REACTION MECHANISM
MALEIC ANHYDRIDE FROM BUTANE

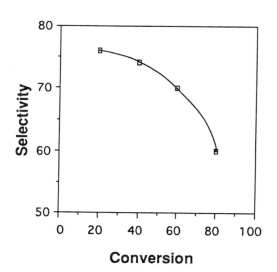

FIG 5. EFFECT OF BUTANE CONVERSION
ON MA SELECTIVITY

selectivity to maleic anhydride really drops to very low levels. Hence, once through processes are operated at conditions which will result in the highest yield, defined as the ratio of the desired product to feed. As an example, the effect of conversion on selectivity for the production of maleic anhydride is shown in Figure 5. A decrease in butane conversion from about 80% to about 20% can increase the selectivity to maleic anhydride by about 16%.

5 TYPICAL PROCESS

A typical partial oxidation process, shown in Figure 6, consists of a reactor containing the appropriate catalyst for the desired reaction. The reactor can be either a fixed or fluidized bed, to which the feed hydrocarbon and the oxidant (typically air) are fed either separately in the case of a fluidized bed reactor or pre-mixed in the case of a fixed bed reactor. Since the feeds are introduced separately into a fluidized bed reactor, the concentrations of hydrocarbons are typically much higher compared to a fixed bed reactor where the hydrocarbon concentration is normally kept below the lower flammability limits. Due to the higher concentration of hydrocarbons in the feed, the productivity of the fluidized bed reactors is substantially higher compared to the fixed bed reactors. However, due to the violent movement of gas and solid inside a fluidized bed reactor, the catalyst has to be designed and formulated to withstand the attrition.

Also, inside a fluidized bed reactor, there is an enormous amount of gas and solid circulation which makes the residence time of a typical fluidized bed reactor higher compared to the average residence. Due to this phenomena called gas back-mixing, there is a substantial amount of product decomposition in the reactor and which reduces the yields achieved by a fluidized bed reactor compared to a fixed bed reactor.

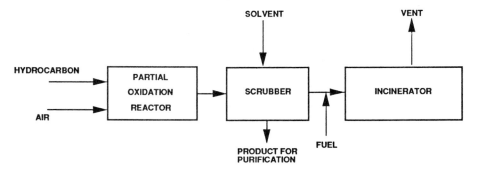

FIG. 6. TYPICAL PARTIAL OXIDATION PROCESS

Recently duPont has developed a new catalyst for the production of maleic anhydride which has enabled them to design a transport bed reactor, which will simulate a fixed bed reactor with respect to gas flow, for the production of maleic anhydride.[10] This strong catalyst has enabled duPont to separate catalyst oxidation and reduction processes, the two steps in the redox process, and achieve a higher selectivity.

In typical oxidation processes, the reactor off-gases are cooled to generate steam and fed into a scrubber in which the products are removed as a solution by contacting it with an appropriate solvent. Water is the commonly used solvent. The removed product is sent to a purification step to produce the desired product. The scrubber off-gases are normally incinerated, most of the time with additional fuel, to meet the environmental regulations.

6 OXYGEN-BASED PROCESS OPTIONS

As discussed earlier, the air-based once through process is optimized to maximize the once through yield. This process suffers from several disadvantages. They include: the process does not operate at the maximum selectivity and the unreacted feed hydrocarbon from the reactor is not recovered but sent to an incinerator. To counter these problems, oxygen-based processes have been developed and the following partial oxidation processes have been switched from air to oxygen:

> production of vinyl chloride from ethene
> production of vinyl acetate from ethene
> production of ethylene oxide from ethene

Also, an oxygen-based process becomes more attractive compared to an air-based process when the process is operated at a higher pressure. Several other air-based processes are being evaluated to use oxygen to improve the overall economics.

As an example, let us look at the production of ethylene oxide from ethene. The process was initially developed based on air as the oxidant. In this process also, as the conversion of ethene increases, the selectivity to ethylene oxide decreases rapidly. In order to achieve a reasonably high overall selectivity and yield, the process was operated using two or three different reactor and scrubber trains in series. The first reactor was operated at a lower conversion to achieve a higher selectivity while the second reactor was operated at higher conversion to decrease the loss of unreacted ethene. Figure 7 shows an air-based ethylene oxide process and this involves two reactors and two scrubbers.

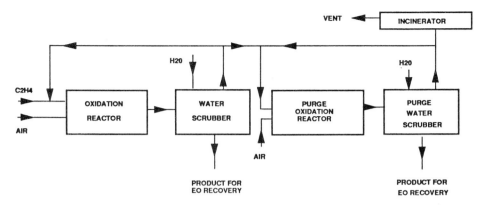

FIG. 7. PRODUCTION OF ETHYLENE OXIDE (EO) USING AIR

Halcon/Scientific Design developed an oxygen-based process[11] in which only one reactor was employed and it was operated at a lower ethene conversion to maximize the selectivity to ethylene oxide. Figure 8 shows an oxygen-based ethylene oxide process. Typical conversion and selectivities for the air-based two reactor process and oxygen-based process are shown in Table 2. From the table one can see that in the air-based EO process, ethene conversion is increased in steps from 20% to 35% to 60% while sacrificing ethylene oxide selectivity whereas in the oxygen-based process, the conversion is kept low to achieve a high selectivity. Most of the recent plants are based on oxygen due to the attractive economics. Generally, oxygen-based processes have been found to be economically attractive for retrofit purposes when the feed hydrocarbon is an olefin due to its high cost. However, when the feed hydrocarbon is a paraffin, air remains as the best option. However, for a new plant, oxygen can be used to reduce the size and capital investment.

Table 2

Process Operating Conditions for the Production of Ethylene Oxide - based on Air and O_2

	Air-Based Process			O_2-based Process
	First Stage	Second Stage	Third Stage	
T, $^{\circ}$C	250-270	250-270	260-280	250-270
C_2H_4 Conv.	20	35	60	15
EO Sel.	72	69	65	78

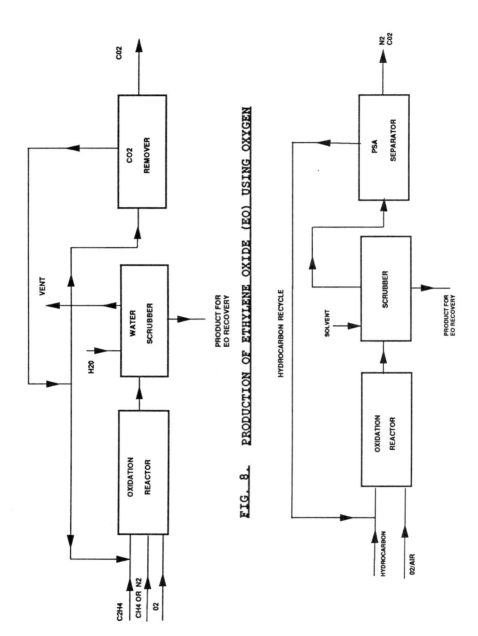

FIG. 8.   PRODUCTION OF ETHYLENE OXIDE (EO) USING OXYGEN

FIG. 9.   BOC PROCESS FOR ETHYLENE OXIDE (EO)

7   NEW PROCESS OPTIONS

The use of pure oxygen, however, brings a variety of issues to the chemical processes. They include: cost of oxygen, materials of construction (this could become a major issue for retrofitting an existing process) and flammability considerations. To address these issues, the use of oxygen enrichment is always considered. As mentioned earlier, the use of pure oxygen as the feed, allows one to operate the reactor at a lower conversion and recycle the unreacted hydrocarbons. However, the presence of large amounts of N_2 in oxygen-enriched air does not allow a direct recycle without a substantial loss in production.

Most of the conventional separation systems such as distillation and absorption are not very good in recovering light hydrocarbons such as C_2's, C_3's and C_4's from dilute and low pressure streams. Since most of the partial oxidation processes operate at low pressures (ethylene oxide production is an exception), conventional separation techniques were found to be economically not attractive for this application. This prevented the use of oxygen-enriched air as the feed for most of the processes.

Recently, new processes, shown in Figure 9, have been developed at BOC Group Technical Center based on pressure swing adsorption (PSA) technology[12] which are particularly effective in removing N_2 from dilute hydrocarbon containing streams. These processes operate efficiently at low pressures (<2 barg) and have been found to be very effective in rejecting N_2 (>90%) while recovering almost all of the hydrocarbons. These PSA processes can also be used to reject CO_2 and hence the need for an additional CO_2 separator can be eliminated. Also, by performing the separations at low pressures, these processes are safer compared to traditional separation processes.

8   Pressure Swing Adsorption

Pressure Swing Adsorption (PSA) process can either be equilibrium or kinetically controlled. In the equilibrium controlled processes, the hydrocarbon is typically adsorbed at the feed pressure and the desorption occurs either at a lower pressure (or under vacuum) or in the presence of an inert gas. In these processes, the change in the partial pressure of the hydrocarbon desorbs the hydrocarbon. In the kinetically controlled processes, the rate of adsorption and desorption are the key phenomenon. In these processes, the inerts may be produced as an adsorbed product while the hydrocarbons will be the non-adsorbed product. Also, depending on the adsorbent, either the hydrocarbons or inert component can be adsorbed and the other can be produced as the non-adsorbed species.

During a PSA process, the following steps are typical: adsorption or feed; pressure equalization to conserve energy; and desorption. Several other steps such as purge; product backfill etc. are usually added to further improve the process. Since PSA processes operate in a semi-continuous mode, buffer tanks are provided for the product to achieve a steady product flow rate at the desired pressure. Alternatively, a multibed system (more than 2 beds) can be used to achieve virtually a continuous production.

9   CONCLUSION

A new process, which enables one to take advantage of the increase in process selectivity by operating the process at a lower conversion, for the production of petrochemicals is presented and is based on pressure swing adsorption technology and allows one to use oxygen enriched air as a feed instead of pure oxygen.

REFERENCES

1.   J. B. Edgerly, Chemical Processing, 1990, December, 21.
2.   W. R. Moser, Catalysis by Organic Reactions, Marcel Dekker Inc., 1981.
3.   A. Bielanski, and J. Haber, Oxygen in Catalysis, Marcel Dekker Inc., 1991.
4.   P. Mars, and D. W. van Krevelan, Chem. Eng. Sci., 1954, 3, 41.
5.   J. R. Monnier, and G. W. Keulks, J. Catal., 1981, 68, 51.
6.   L. D. Krenzke, and G. W. Keulks, J. Catal., 1980, 64, 295.
7.   G. W. Keulks, Z. Yu, and L. D. Krenzke, J. Catal., 1983, 84, 38.
8.   R. M. Contractor, and A. W. Sleight, Catal. Today, 1987, 1, 587.
9.   J. C. Burnett, R. A. Keppel, and J. D. Robinson, Catal. Today, 1987, 1, 537.
10. R. M. Contractor, H. E. Bergna, H. S. Horowitz, C. M. Blackstone, B. Malone, C. C. Torardi, B. Griffiths, U. Chowdhry, and A. W. Sleight, Catal. Today, 1987, 1, 49.
11. R. A. Meyers, Handbook of Chemicals Production Processes, McGraw-Hill Book Co., 1986.
12. R. Ramachandran, Y. Shukla, and D. L. MacLean, U.S. Patent 4,868,330; 4,987,239.

Synthesis Gas Production by Convective Reforming

J. F. G. Ellis[1], A. D. Maunder[1], and A. Czimczik[2]
[1] BP ENGINEERING, I HAREFIELD ROAD, UXBRIDGE UB8 IPD, UK
[2] L&C STEINMÜLLER, GUMMERSBACH, GERMANY

1   INTRODUCTION

The Priestley Conferences provide an ideal opportunity for significant new developments in oxygen use to be described.   The development outlined in this paper has resulted from combining BP's requirements and ideas for a low cost route for production of Synthesis Gas (Syngas) with L&C Steinmüller's proven mechanical engineering, fabrication and prototype testing experience.

Over the last few years considerable developments have been achieved by the process industry in the reduction of the energy requirement for Syngas production.   One of the key reasons for this reduction has been the cost efficient use of oxygen.   Convective Reforming, as described in this paper, provides an opportunity for reducing syngas costs still further.   The developments which we describe can represent a considerable improvement relative to commercial processes currently available.

Other convective reforming developments have been undertaken over the years by many companies e.g. ICI, Uhde, Chiyoda, Stone & Webster and others, but as yet no full scale commercial plants using oxygen have been built.   Space only allows for a short description of what has gone before.

1.1 Syngas

Syngas consists of mixtures of gases primarily containing hydrogen and carbon monoxide, and in some case nitrogen. Syngas production is mature technology having been practised in one form or another with various feedstocks for the last 60 years.   Syngas is an intermediate for Ammonia, Methanol, Fischer-Tropsch hydrocarbons, OXO alcohols, acetyls (acetic acid/anhydride) and hydrogen.   It is also used as a reducing gas for metallic ores, for example iron.   Syngas may also be considered as a basic intermediate building block which provides a means by which oxygen may often be incorporated into chemical products.

For economic reasons, the use of natural gas as a feedstock for syngas has become more prevalent in recent years. Existing processes for syngas production, for instance tubular steam reforming, fixed bed autothermal reforming (using oxygen or air) and non-catalytic partial oxidation, having been in operation for many years, are well known. The reactions for steam reforming [1] and oxygen reforming/partial oxidation [2] are given below :-

$$CH_4 + H_2O \quad = \quad 3 \ H_2 + CO \ \ldots\ldots [1]$$

$$CH_4 + 1/2 \ O_2 \quad = \quad 2 \ H_2 + CO \ \ldots\ldots [2]$$

The most commonly used syngas producer in the past has been the steam reformer. This is a unit of large dimensions, requiring many ancillary systems such as steam generation, fuel and combustion air handling. The physical sizes and capital costs of syngas producers using oxygen are much lower than that required for steam reforming. This improvement is currently offset by the cost of providing oxygen as this is a major contributor to the overall production cost. However, none of the above processes is dominant.

The capital cost for syngas production, when this is required in an overall project making one of the products mentioned above, is a large proportion of total cost. In the case of ammonia or methanol, this may contribute more than two thirds of the total capital required. The incentive for new syngas process developments has thus been to reduce cost typically by simplification, by reduction in size and by the utilisation, where possible, of new materials. Oxygen processes have been much improved over the past few years, especially in the context of stand alone plant. Nevertheless, a major reduction in syngas costs would result from the development of a really low cost source of oxygen, either on a stand alone basis or perhaps by integration with the syngas production process.

1.2 Combined Reforming

Combined Reforming produces synthesis gas having a stoichiometric content of hydrogen to carbon oxides optimum for methanol synthesis, when using natural gas feedstock. This is achieved by a combination of an externally heated primary reformer in series with an oxygen fed autothermal secondary reformer (Figure 1). As the product gas has the desired stoichiometric composition for methanol production, methanol synthesis efficiency is increased.

With this system it is possible to operate the primary reformer at an outlet temperature of 750-780 °C, compared with 880 °C typical of the conventional steam reforming methanol plants. This permits operation of the process at 35-40 bar instead of the 20 bar of conventional plants, with a consequent saving in synthesis gas compression costs. The reduction in operating temperature improves plant reliability.

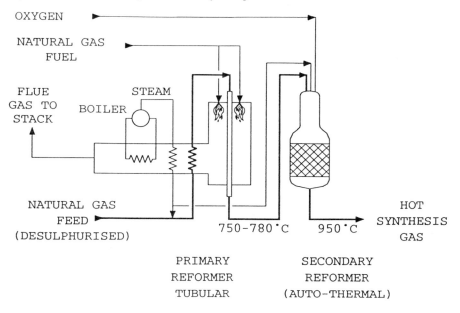

OXYGEN

NATURAL GAS
    FUEL

FLUE
GAS TO
STACK

BOILER

STEAM

NATURAL GAS
    FEED
(DESULPHURISED)

750-780°C   950°C

HOT
SYNTHESIS
GAS

PRIMARY
REFORMER
TUBULAR

SECONDARY
REFORMER
(AUTO-THERMAL)

Figure 1 : Combined Reforming

Due to the autothermal component of the overall reforming process, the heat transfer duty of the primary reformer is about 25% of that in the conventional single stage steam reforming process.

1.3 Convective Reforming

The Convective Reforming process which BP and Steinmüller have worked on is a particular type of Combined Reforming. It comprises two-stage combined convective/autothermal reforming (Figure 2). Pre-heated natural gas and steam are reacted over steam reforming catalyst, but this is done in a convectively heated primary reformer, rather than a radiant unit as described previously. The convective unit would have a much smaller physical size than the comparable radiant unit. The hot product gases from the secondary reformer are cooled to provide the necessary reaction heat to the primary reformer rather than being used to raise steam.

The theoretical advantages of this type of operation over combined radiant/autothermal reforming are as follows:-

• use of process heat for the reforming duty reduces the size of, or in some situations eliminates the need for, steam raising plant which is necessary with a radiant reformer.

• the high temperature tubes in this type of convective

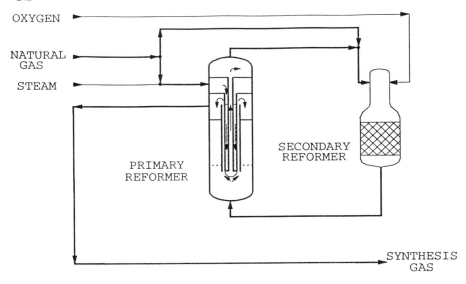

OXYGEN

NATURAL
GAS

STEAM

PRIMARY
REFORMER

SECONDARY
REFORMER

SYNTHESIS
GAS

Figure 2 : Convective/Autothermal Reforming

reformer only need to withstand a relatively small
differential pressure caused by a pressure drop around
part of the system, rather than the total system pressure
which occurs with a radiant unit. This enables the use
of thinner tubes with the potential for longer life. A
much higher syngas generation pressure can also be used,
even greater than the already increased value for
Combined Reforming, perhaps up to 50-60 bars.

• the convective unit is much more compact and lighter in
  weight than the comparable radiant unit.

    In the process, some of the desulphurised fresh feed gas
is combined with HP steam and pre-heated before entering the
primary side of the convective reformer. Here, the gas
passes down the catalyst filled annuli of the reformer tubes,
and heat is transferred across the tube walls from hot syngas.
The partially reformed gas leaves the tubes and passes back
up the central return tubes leaving the CR primary side at
about 700 °C. This gas is combined with the remaining
natural gas and fed with pre-heated oxygen (200 °C) to the
autothermal reformer burner. The gas is partially oxidised
in the combustion zone of the autothermal reformer and then
passes through a fixed bed of reforming catalyst. Oxygen
feed rate is such that syngas exits the autothermal reformer
at 950-1050 °C. This effluent syngas passes back to the
shell side of the Convective Reformer where guide tubes
direct its flow along the outside of the reformer tubes,

maintaining a velocity which enables good heat transfer.
Heat is exchanged across the tube walls, cooling the
autothermal reformer effluent gas to about 650 °C.

2. CONVECTIVE REFORMER DESIGN

BP's developments described here have been in
collaboration with L&C Steinmüller GmbH. Steinmüller have
previously designed and fabricated a very successful
prototype helium heated convective reformer for KFA Jülich.

The 18 tube unit was tested at KFA Jülich with
electrically heated helium as the shell side gas. It
transmitted about 5 MW and operated for approximately four
thousand hours. In the last operating run it was subjected
to intensive temperature cycling. It may be seen from Figure
3 that the temperature of the inlet helium was cycled between
620 °C and 950 °C one hundred times in a period of about 20
days. The fact that no mechanical deterioration occurred
emphasises the integrity of the reformer and the soundness of
the fundamental design concept.

Although this unit did not use oxygen, it operated at
temperatures very close to that required for the oxygen
process and hence represented an ideal starting point for the
joint BP/Steinmüller development effort.

3.1 Mechanical Design

The hottest components in all tubular steam reformers

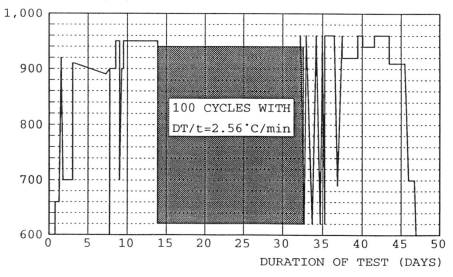

Figure 3 : Temperature Cycling Profile of 18 Tube Bundle

operate at temperatures which are close to the maximum
allowable for Nickel based alloys.   Given the economic
requirement for a tube operational life of 100,000 hours,
satisfactory performance can only be achieved by minimising
both the applied stress and the number of stress
applications.   There are four key factors which enable the
former to be achieved; the latter is concerned with
steadiness of operation.

      As described before, the reformer tubes, of which there
may be 100-200 in a commercial design (Figure 4), operate
with only a **small differential pressure** (of approximately 2 -
5 bar).   Thus, whatever the absolute pressure, the pressure
stress in the tube is small and results only from pressure
drop through the system.   The reformer tubes are 120 mm
inside diameter and 6 mm thick.   By contrast, in conventional
reformers the tubes are subject to the absolute pressure of
the feed and may be as much as 20 mm or more in thickness.
Apart from the expense, the thicker tube is much more
vulnerable to thermally induced stress when temperature
fluctuates for instance during start up and shut down.

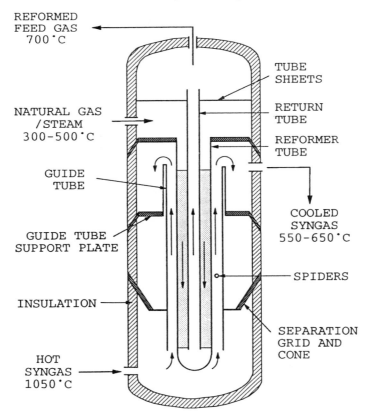

Figure 4 : Convective Reformer Arrangement

The arrangement in which the **reformer tube is held concentric within a guide tube by spiders** (Figure 4) is the second feature. This ensures that the reformer tube is subject only to controlled axial flow.

Thirdly, by using **guide tubes** it is ensured that there is no cross-flow and minimal flow perturbation; the former would cause thermal stress and distortion, and the latter could lead to fatigue.

Fourthly, by using a **return tube to take the reformed gas away from the catalyst**, the axial thermal expansion of the reformer tube is totally unrestrained. Both reformer and return tubes can freely expand downwards. The design has another important feature in that each tube assembly is able to move radially and longitudinally independently of the others and without any undue loads being imposed.

The shell of the reformer is made from Cr/Mo steel which is internally lagged with fibre insulation. The internals are supported on two cones, which pass through the insulation and accommodate differential expansion. The reformer tubes are made in a special alloy and the other high temperature components are in Alloy 617 or Alloy 800.

A theoretical comparison has been made of the creep performance of the convective reformer tubes as against those in a radiant reformer. Inelastic finite element models of the tubes were prepared, and the thermal and pressure stresses calculated in each case. The results of these calculations indicated that the required 100,000 hour tube life can be achieved.

3.2 Carbon Related Issues

There are various issues requiring consideration in order to prevent carbon related problems occurring when cooling the secondary reformer reformer product gas. The design of the Convective Reformer ensures that both the primary and autothermal reformer catalysts are operating in regimes which are thermodynamically "safe" with respect to carbon formation (via equation), the Boudouard reaction, (see Reference 1 for a more detailed discussion of catalyst coking in steam reformers). However, a key area of the Convective Reformer design is the annular gap between the outside of the reformer tube and the inside of the guide tube in which the autothermal reformer effluent transfers its heat into the primary reformer. This has the highest potential tendency for carbon-related corrosion of the reactor materials. This stems from the high carbon monoxide content of the gas mixture in this region.

$$2\ CO(g)\ =\ CO_2\ (g)\ +\ C\,(s) \dots\dots [3]$$

The Boudouard reaction [3] is thermodynamically favoured at high carbon monoxide/carbon dioxide ratios, at low temperatures and higher pressures. This means that the

potential for carbon corrosion is greater for methanol
(oxygen fed) than for ammonia (air fed) applications, and
also varies along the annular gap as the gas temperature
changes. Figure 5 shows three different processes which can
affect the mechanical integrity of the alloy tubes forming
the annulus :-

Carbon Deposition is the physical formation of carbon on
the surface of the alloy tube. This can occur at
temperatures where the Boudouard reaction is favourable and
where a catalytically active metal surface is exposed. It is
known that elements such as iron or nickel, which are present
in most high temperature alloys, can catalyse the Boudouard
reaction and result in the formation of highly characteristic
filaments or whiskers of carbon. The mechanism of filament
growth keeps the catalytic metal exposed to the gas at the
head of the growing whisker. These filaments can grow to
great lengths and can lead to partial or total blockage of
the annular gap as well as decreased heat transfer
efficiency.

Carburisation is a high temperature process in which
carbon diffuses causing consequent loss of mechanical
properties. The autothermal reformer product is at
sufficiently high temperature that the Boudouard reaction
(Equation 3) is not favoured and the driving force for
carburisation is relatively low in the high temperature
regime. As the gas cools the driving force increases but
carbon diffusion rates are slow below about $850\,^\circ$C, so that
carburisation does not represent a significant risk.

Metal Dusting is closely related to carbon deposition but
is a phenomenon in which extensive pitting of metal surfaces
occurs rapidly accompanied by significant carbon growth

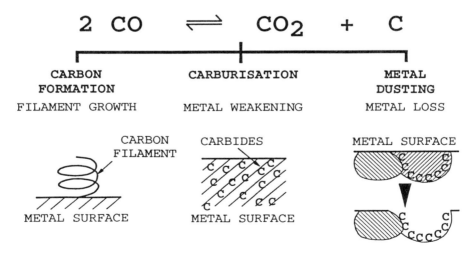

Figure 5 : Carbon Corrosion Processes

within the pits. The magnitude of the pitting is thus much
greater than the very small levels of metal loss associated
with "normal" carbon deposition. The rapidity of metal
thinning and carbon growth with metal dusting could lead to
severe operational problems. BP work has shown that the
corrosion debris contains no original alloy but consists
entirely of a mixed metal oxide spinel phase and metallic
crystallites/carbon filaments, indicating the role that both
carbon growth and oxidation play in driving metal dusting.

3.3 Materials Testing

The selection of materials of construction of the
Convective Reformer, and the heat transfer tube in
particular, is obviously key to its development. In addition
to high temperature mechanical strength, the materials must
be resistant to the three carbon corrosion processes outlined
above. In order to carry out this selection an extensive
test programme has been carried out at BP Research, Sunbury,
UK.

The programme involved preliminary screening tests using
simple atmospheric pressure apparatus and thermogravimetric
techniques to identify a series of preferred candidate
materials. These materials were further tested in purpose-
built high pressure apparatus with the capability to control
gas composition, temperature, pressure and gas velocity
independently, and which simulate conditions in the heat
transfer annulus. Test sections of alloy up to 0.5 m length
were tested for several weeks. This apparatus was used to
gain information on metal dusting and carbon formation, while
long term carburisation tests were carried out under a range
of conditions. Carbon deposition was monitored by continuous
pressure drop measurements and post-test by extensive use of
analytical techniques, in particular electron microscopy.

As a result of this extensive programme, a more detailed
understanding of the kinetic and thermodynamic factors at the
gas/alloy interface has been accumulated. Using this,
materials have been selected which are resistant to
carburisation, carbon formation and metal dusting under the
most severe potential operating conditions.

3.4 Thermal/Gas Flow Design

Thermal design and optimisation of the convective
reformer configuration were carried out using a computer
model jointly developed by BP and Steinmüller, which was
calibrated to the performance of the 18 tube prototype unit
at KFA Jülich.

Kinetic calculations for the steam reforming and carbon
monoxide shift reactions were carried out stepwise through
the catalyst along the reformer tube, in tandem with heat
transfer calculations. Use of the model has allowed us to
develop an optimised design configuration, and to evaluate
the effects of tube wall fouling, catalyst ageing and

poisoning. An example of the predicted temperature profile
is shown in Figure 6.

The modular design of the tube assembly with its external
guide tube implies little tube-to-tube variation during plant
operation. This is a source of confidence when predicting
thermal performance at larger scale. Second order scale-up
effects have been investigated using state-of-art
computational fluid dynamics (CFD) tools to optimise the hot
syngas inlet system at the bottom of the Convective Reformer.
This enables design configurations to be checked so that
constant temperature and flowrate to each tube, taking into
account dimensional tolerances, can be achieved. Cold flow
modelling of this system has also been carried out.

TEMPERATURE ('C)

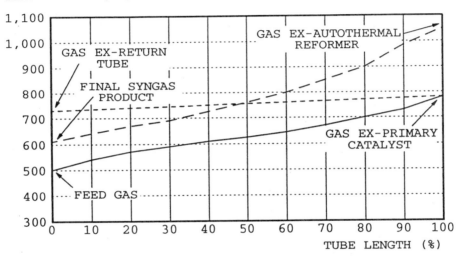

Figure 6 : Computer Predicted Temperature Profiles

4.   PROCESS OPTIONS

4.1 Methanol Production

Methanol syngas generation is carried out using the
Convective Reformer, as previously described, combined with a
commercially available autothermal reformer.

Using natural gas, the process can produce syngas with a
stoichiometric ratio :-

$$\frac{H_2 - CO_2}{CO + CO_2} = 2.04 \quad \ldots\ldots\ldots\ldots [4]$$

as feed to the methanol synthesis area. The exact ratio from reforming is achieved by varying the bypass of fresh natural gas around the Convective Reformer. A further consideration is that reasonable temperature approaches must be selected across the Convective Reformer exchanger so that it is of economic size.

Any of the commercially available synthesis processes can be used for synthesising the methanol and purifying it.

The economics of this process for methanol have been studied :-

Regarding capital cost, because of the autothermal component of the overall reforming process, the heat transfer duty of the primary reformer in combined reforming, whether convective or radiant, is about 25% of that in the conventional single stage reforming process. This, in either case, greatly reduces the size and cost of the primary reformer. Of course, this saving compared to single stage reforming is offset by the cost of supplying pure oxygen, as the Air Separation Unit providing the oxygen for a commercial scale plant accounts for up to 25% of the total capital cost. The differential between combined and single stage reforming basically depends on a relatively small difference in two large quantities, the capital cost of a steam reformer and that of an oxygen plant. As these items are generally obtained from different sources, these costs are subject to varying commercial factors. Differential utilities costs may also be significant, as these depend on the type of drives (steam or gas turbine or electric motor) used for the oxygen plant and the availability or non-availability of an external (constant or emergency) energy supply.

Convective Reforming has the additional advantage that the capex required for a Convective Reformer is 30% less than that for the comparable Radiant Reformer. If space is at a premium on an existing site, Convective Reforming with its compact size could well be even more favoured. This factor could be particularly important in the offshore context where size and weight are both crucial. It must noted, however, that the question of oxygen supply offshore remains to be addressed.

The operating cost advantage for convective reforming will depend on feed and fuel costs. The combined reforming process for methanol using a radiant primary unit has an overall energy consumption 5-10% less than the best single stage process. Convective Reforming will allow further optimisation of this process system. Because of its relative simplicity, Convective Reforming will also require much lower maintenance and manning levels than the conventional steam reforming step used as the first stage of combined reforming.

4.2 Carbon Monoxide Production

A compact and efficient convective reforming route for

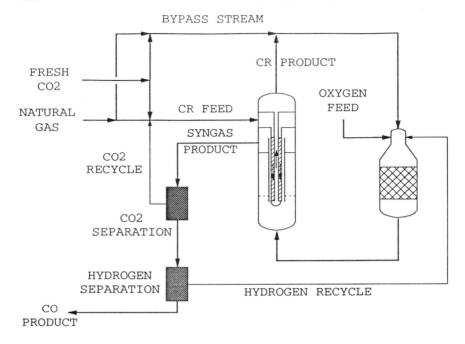

Figure 7 : Carbon Monoxide Production

carbon monoxide production (Figure 7) has been outlined in a
recent publication (Reference 2).

The proposed scheme uses the same two stage reforming
configuration as previously described, but with the important
difference that carbon dioxide, rather than steam, is added
to the primary reformer feed. The energy requirement of the
convective reforming process is covered by the fuel energy in
the methane feedstock. The only waste product is water. No
carbon dioxide at all is released as this is removed from the
product gas by amine scrubbing and is recycled to the
reformer. Carbon monoxide is separated from the hydrogen in
a cold box. If there is a local requirement for hydrogen, it
can be purified and supplied directly. If there is no
requirement for hydrogen, the cold box tail gases, containing
hydrogen and some carbon monoxide, can be returned to the
process. This saves significant quantities of methane since
further carbon monoxide is produced by direct reaction of
hydrogen with carbon dioxide.

Besides the reduction in production costs for each tonne
of carbon monoxide, the low environmental impact of this
process must be seen as a positive advantage.

4.3 Other Process Options

Convective reforming is well suited to the preparation of
synthesis gas for **ammonia production** with relatively little

modification being required to the existing flowsheet concept. Ammonia production is normally carried out by Combined Reforming, but with the secondary reformer fed with air, not oxygen. However, when Convective Reforming is used, oxygen or oxygen enriched air may advantageously be considered as, if this is not done, the maximum heat recovery in the Convective Reformer is usually not sufficient to match the primary reforming heat load when a stoichiometric gas is required.

Convective Reforming can be used for **Natural Gas Conversion** to hydrocarbon liquid transport fuels, either via methanol and the Methanol to Gasoline (MTG) process or by converting the syngas directly to distillate hydrocarbons via the Fischer-Tropsch (FT) process. In either case, the use of Convective Reforming would be, as for methanol alone, a major contributory factor in the process. For Gas Conversion, parts of the process could be much greater in scale than for methanol; in some areas, the capacity per train might be comparable. The economics here are affected by the route to be employed to convert the synthesis gas to the final product, for instance whether to Fischer-Tropsch liquids such as Middle Distillates or to Gasoline.

5. ENVIRONMENTAL BENEFITS

Environmental issues are of increasing importance today. The balance between capital expenditure and the potential environmental impacts have never been so critical.

In the case of the Convective Reforming process as described above, the recuperative nature of the process enables the high grade heat of the synthesis gas to be recycled into the system, so reducing the need for the firing of fuel. The environmental advantages are clearly those of reduced emissions, for example carbon dioxide, NOx, etc., and less wasted energy.

Carbon taxes, related to the amount of carbon dioxide produced by individual processes have been introduced in Scandinavia. Where carbon dioxide is suitable as a feedstock, a further reduction in emissions can be achieved. A process, such as carbon monoxide production mentioned earlier, which is a net carbon dioxide consumer is part of a possible future set of processes which reduce atmospheric emissions and pollution by using waste products as a feedstock.

6. CONVECTIVE REFORMER DEMONSTRATION

As can be seen from the account above, a large number of issues in the development of this technology have already been addressed and satisfactorily resolved. However, it is still regarded as necessary to erect a small demonstration unit before proceeding to a full scale commercial plant. This is in order to confirm that there will be no unresolved carbon related problems under identical process conditions to those anticipated in the commercial unit. The other, secondary, point would be to assess operability and identify the best method of control under normal operation and transient conditions. It would also provide process data to enable optimisation of the commercial plant design.

BP and L&C Steinmüller have already jointly completed the process and mechanical design of a Single Tube Experimental Plant (STEP) which is based upon a commercial size tube assembly.

As mentioned before, each tube assembly of the 150 tubes typically in a commercial scale convective reforming plant effectively performs as a separate module. The feed and syngas flows to each tube are identical and independent of the other tubes. Realistic process demonstration can therefore be achieved using a single tube assembly having commercial scale dimensions. When a single tube is fed with gas at commercial plant feed conditions the gas velocities and physical properties will be the same as in a commercial reactor. The heat transfer characteristics and the potential for carbon deposition in the annulus will also be identical. It is therefore unnecessary from both the process and metallurgical standpoints to operate with more than one tube, and the results from the STEP programme can be directly transferred to commercial plant designs.

The demonstration plant to prove the process concept has been designed for a number of potential applications. The test programme includes two key objectives. Most importantly, confirmation will be obtained of the lowest steam:carbon ratio that can be safely used with the selected materials, consistent with minimal carbon deposition under realistic operating conditions. Secondly, the temperature in the autothermal reformer and hence the oxygen flow to the burner, will be optimised.

The mechanical design of the demonstration convective reformer will, so far as possible, be generally similar to a commercial unit. Particular attention has been paid to features such as the attachment of the reformer tube to the tubesheet, the internal and external thermal insulation and the removal and replacement of the tube assembly.

The proposed demonstration plant includes electric gas pre-heaters, a natural gas desulphuriser, a single tube convective reformer, an autothermal reformer and syngas product coolers. The entire plant has been designed as a road transportable module having a "footprint" of 4 by 5 metres.

Some preliminary design work has also been carried out for a single tube demonstration plant, based on a single tube of reduced scale. This will be able to reproduce some of the essential features of a commercial plant but not all.

9. CONCLUSION

It is believed that the developments described in this paper offer an environmentally attractive and cost effective route for syngas production through the use of oxygen. Clearly, the challenges which remain to be addressed, in particular the successful process demonstration under optimal conditions, are ones in which perseverance and strength of purpose are required. The benefits of moving ahead progressively in this technology are that cost effective applications may be found for syngas based processes which might hitherto have been unattractive. Various arrangements are in place between BP and L&C Steinmüller on the commercialisation of this technology.

A further challenge must be to reduce the cost of tonnage oxygen because a large part of the cost of any of these process options is attributable to the cost of supplying oxygen. Insofar as this has been a continuing preoccupation of the oxygen suppliers to all processes, this objective is not new. However, the reduction of oxygen cost in the specific context of lower cost syngas production may represent the next advance. It is to be hoped that opportunities might emerge for integration of reforming with oxygen supply.

As a final comment, it is interesting to note that although syngas production might appear to be a mature technology, opportunity exists for further development over the next 20 years or so. The type of developments described in this paper will be a part of this as indeed will be the application of novel materials together with enhancements in the supply of oxygen.

10. ACKNOWLEDGEMENTS

The authors would like the thank many colleagues in the Steinmüller and BP organisations. In particular thanks are due to the many people in Steinmüller in the knowledge that this development is founded upon their previous work. We would also like to thank those in BP Research who carried out very valuable work on carbon formation and contributed the relevant section of the paper, and to many other colleagues in BP Engineering who have likewise done the essential work on Process, Mechanical and Thermal design, and have also contributed to this text.

11. REFERENCES

1. Steam Reforming Analysed, E.S. Wagner, G.F. Froment, Hydrocarbon Processing, 1992 (July), page 69-77

2. Convective reforming, A. Czimczik, R.P. Manning, Hydrocarbon Technology International, 1992, page 59-64

Wet Oxidation and Biological Processes

Use of Oxygen in the Pulp and Paper Industry

Ingemar Croon

CROONCONSULT, BOX 135, 182 05 DJURSHOLM, SWEDEN

1 INTRODUCTION

Annual consumption of paper worldwide amounts to about 260 million tons. The ratio in the processing of the fiber raw material is 40 percent chemical wood pulp, 18 percent high-yield wood pulp, 32 percent recycled fiber, 8 percent annual fiber (straw) and 2 percent miscellaneous. The kraft process is dominant in the production of chemical wood pulp fibers, with both soft and hardwoods being used.

Wood is a highly complex substrate, containing several hundred chemical entities. The main constituents are:

Cellulose (40-45%) is a linear polymer with a degree of polymerisation of about 10.000 glucose units (compared with 1.000-2.000 in a strong pulp and 200-300 in a weak pulp)

Hemicellulose (20-25%) are non-linear polymers built up by various sugars (xylose, glucose, mannose, galactose etc.) with a degree of polymerisation of 200-300.

Lignin (20-30%) constitutes the glue between the fibers and is a three-dimensional polymer built up by various aryl-alkyl monomers.

Resinous compounds (2-5%) fatty acids and esters, resin acids and esters, sterols, terpenes and phenolic compounds.

Sugars and salts (2%).

In the kraft process, wood chips are cooked in a liquor of sodium hydroxide and sodium sulphide. To obtain a clean, white pulp, the pulp is subjected to multi-stage bleaching.

The chemicals conventionally used in the bleaching process are chlorine, sodium hypochlorite, chlorine dioxide and sodium hydroxide. Oxygen began being substituted for chlorine and hypochlorite in 1975, a trend which has since accelerated.

Figure 1 illustrates a typical, traditional bleaching sequence used in processing kraft softwood pulp, and a comparison with modern bleaching sequences.

1970	1990	1995
Chlorination	Oxygen	Oxygen
Alkaline extraction	Chlorine dioxide	Ozone
Chlorine dioxide	Oxygen	Oxygen
Alkaline extraction	Peroxide and/or	Peroxide and/or
Chlorine dioxide	Chlorine dioxide	Chlorine dioxide

Figure 1   Old and Modern Bleaching Sequences

During the 1930s, German, French and Swedish researchers studied the interaction between unbleached pulp and oxygen. They discovered that oxygen, in alkaline media, reacted to dissolve lignin. An undesired side effect was that the carbohydrates were degraded with the result that the pulp fibers were damaged, forming a paper of very low strength.

In the Soviet Union (where a tremendous amount of basic research has been performed on wood and pulping chemistry), Nikitin and associates discovered in the late 1950s that addition of magnesium carbonate, prior to the pulp being treated with oxygen in a sodium hydroxide suspension, substantially reduced cellulose degradation.

These findings were the basis of the development work initiated in France, South Africa, Sweden and Canada. Two separate groups were formed, including pulp makers, equipment suppliers and oxygen manufacturers: one group adopted the Sapoxal method (Air Liquide, Sappi and Kamyr), with the other group pursuing the MoDoCIL method (MoDo, CIL and Sunds). In the 1970s, the two groups formed a joint know-how pool, with users licensed to access the pool. Since then, the use of oxygen for kraft-pulp delignification prior to the final brightening stage has gained increasing acceptance (Figure 2).

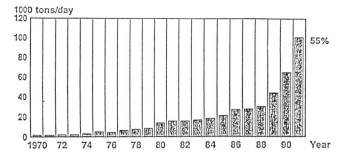

<u>Figure 2</u> Pulp Production Using Oxygen Treatment

Today, approximately 35 million tons of pulp per year is
oxygen-treated. The pulp industry consumes for
bleaching purposes more than 700,000 tons of oxygen
annually. The conditions during the oxygen stage are
shown in Figure 3.

Pulp consistancy	10-12% (or 25-30%)
Oxygen charge	1-2%
Temperature	90-120° C
Time	30-60 min.

<u>Figure 3</u> Conditions in the Oxygen Stage

A stable three phase system, fiber - liquid - gas, is
critical for a good result. Perfect mixing and avoidance
of gas separation are very important. High consistency
(30% fiber) oxygen methods were first introduced but
during the 1970s high shear force mixers were developed
and then medium (10-12%) consistency methods took over.
For ozone treatment the mixing and gas separation factors
are even more important.

2 FROM CHLORINE- TO OXYGEN-BASED CHEMICALS

The introduction of oxygen to the pulping process was the
first step in attaining pollution-free pulp mills. For more
than 70 years, chlorine-based chemicals have been dominant

in pulp bleaching. Chlorinated organic compounds were being discharged by mills into surrounding recipients. The reuse of process water -- long an accepted practice in the industry -- was complicated by the corrosive properties of the chlorine. Thus, the diminished use of chlorine became a central objective. In fact, there are no mills in Sweden or Germany that presently use chlorine. However, chlorination of pulp still remains a common practice in America and Asia, although many mills even there have adopted oxygen technology.

During the alkaline extraction stage, it has been proved that oxygen upgrades efficiency in dissolving residual lignin. For about 10 years, this has been a widely accepted practice within the global pulp industry.

Chlorine dioxide is a gentle-acting and highly efficient brightening agent used in most mills since its introduction in the late 1940s. Due to pressure from environmentalists, combined with improved know-how in using hydrogen peroxide for the same purpose, the use of chlorine dioxide will also decline.

Ozone is the latest chemical used as an oxygen supplement in the bleaching of pulp. Several pulp mills are presently installing facilities for the use of ozone. A bleaching sequence in which an oxygen step is followed by ozone treatment (approximately 0.2 to 0.5 percent ratio to pulp), followed by a second oxygen stage and then by hydrogen peroxide (totally free of chlorine chemicals for bleaching), or a chlorine dioxide stage, is the latest approach to environmentally compatible bleaching.

The life cycles of the most common bleaching chemicals are presented in Figure 4. During the next 10 to 20 years, the pace-setters will be oxygen, ozone and hydrogen peroxide.

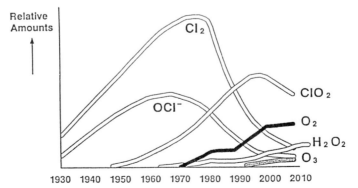

Figure 4   Life Cycles for Bleaching Chemicals

3 REACTION MECHANISM

As previously mentioned, the compositions of wood and pulp are highly complex, with several hundred chemical structures to consider when studying chemical reactions during cooking and bleaching. Also mentioned previously, delignification during pulp bleaching normally occurs during several stages to attain the most selective possible reactions, such as the removal of lignin products without degrading the carbohydrates (cellulose and hemicellulose). A summary of reaction types is provided in Figures 5a and b. The complex course of bleaching can be rationalized considerably if the bleaching reactions are divided into three categories with respect to the initial reacting species.[1]

1. Cationic processes in acidic media (e.g. Cl^+, HO^+, O_3)

2. Radical processes operating in acidic, neutral or alkaline media (e.g. $\cdot O_2^-$ or $ClO_2\cdot$ and secondary radicals as $HO\cdot$, and $Cl\cdot$)

3. Anionic processes operating in alkaline media (e.g. CLO^- or HOO^-)

Figure 5a   Overview of Reaction Types

cationic	radical			anionic
electrophiles	nucleophiles			
oxidative	reductive			
acidic	acidic or alkaline			alkaline
Cl^+, HO^+ H^+	$ClO_2\cdot$, $\cdot O_2^-$, $Cl\cdot$, $HO\cdot$, $HO_2\cdot$	$S_2O_4^{2-} \rightleftharpoons$ SO_2^-		ClO^-, HOO^-
$-2\,e^-$	$-1\,e^-$	$+1\,e^-$		$+2\,e^-$
aromatic and olefinic structures	carbonyl and conjugated carbonyl structures			

Figure 5b   Overview Bleaching reagents
Initially reacting species

Cationic and oxidative radical processes are electrophilic, preferentially attacking electron-rich aromatic and
olefinic structures (such as lignin and resinous extrac-
tives). However, due to their great reactivity, radicals
usually attack both lignin and carbohydrate components. Such
properties limit the selectivity of oxygen bleaching
processes and necessitate multi-stage procedures.

In particular, hydroxyl radicals, stemming from metal
ion-catalyzed dismutation of hydrogen peroxide or by homolytic cleavage of hydroperoxide intermediates, play an
important part in oxygen, hydrogen peroxide and ozone
bleaching processes. Figure 6 illustrates the main routes to
the formation of hydroxyl radicals.

$$M^{n+} + H_2O_2 \longrightarrow M^{n+1} + HO\cdot + HO^-$$

$$M^{n+1} + HO_2^- \longrightarrow M^{n+} + O_2^-\cdot + H^+$$

$$ROOH \longrightarrow RO\cdot + \cdot OH$$

M = metal ion

Figure 6   Examples of Routes to Formation of
            Hydroxyl Radicals

The hydroxyl radical is, in fact, the strongest one
electron oxidant available in aqueous media[2]. Its strong
electrophilic character causes it to react with a variety of
organic and inorganic compounds[3]. Hydroxyl radicals readily
combine with aromatic structures in lignin (Figure 7),
initiating a cleavage of aryl alkyl ether bonds and, under
favourable conditions, the lignin radical formed may react
with molecular oxygen, resulting in complete degradation.

The reaction with cellulose involves the removal of
carbon-bound hydrogen atoms, resulting in the formation of
hydroxyalkyl radicals. These are then converted into the
corresponding carbonyl structures by oxygen treatment, which
initiates peeling, or alkaline degradation of the cellulose
through monomer-by-monomer elimination of glucose. This
describes in brief the adverse effect hydroxyl radicals have
on yield and strength properties of bleached pulps.

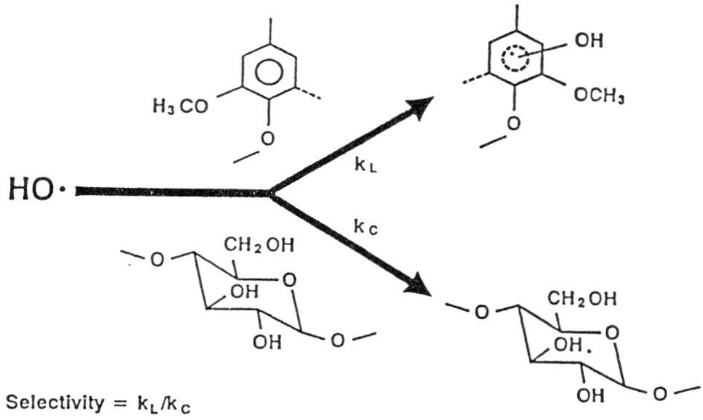

Selectivity = k_L/k_c

Figure 7   Reaction of Hydroxyl Radicals with Aromatic
Structures (Lignin) and Carbohydrates
(Cellulose)

It remains to be determined what role the superoxide
anion $O_2^-\cdot$ (see Figure 6) plays in the highly complex and
numerous reaction variables of pulp bleaching. The nucleophilic reactions seems to dominate over the electrophilic
ones. A great deal more remains to be learned about
controlled applications of oxygen.

As mentioned, heavy metal ions and their quantity must
be closely controlled. This is accomplished mainly through
the use of magnesium salts. The metals are blocked through
the formation of complexes with magnesium hydroxyl compounds.
In certain cases, the pulp is subjected to an acid wash pri-
or to oxygen bleaching. Complexing agents such as EDTA are
also gaining acceptance.

4 MISCELLANEOUS

The technology involved in the biological treatment of mill
effluents is widespread and is mainly directed toward lowering the biological oxygen demand of the effluent. The
application of atmospheric air is most common but certain
processes use oxygen. The latter are more compact and greatly reduce unpleasant odors in surrounding areas. As the use
of oxygen becomes more widespread and on-site facilities for
producing oxygen become common, oxygen-based treatment of
effluents will probably gain wide acceptance.

Oxygen-enriched air can be used to increase the
capacity of the recovery boilers and lime kilns. Another
oxygen application is oxidation of hydrogen sulfide in
cooking white liquor to make it useful as alkali in oxygen
delignification.

5 FUTURE DEVELOPMENT

Interest is already focused on oxygen as an important
chemical in the kraft pulp industry. Indications clearly
point to its increased use. This is especially likely as
mills convert to closed water cycle systems, that is, using
very little fresh water (2-5 m^3 of water is required per
ton of pulp produced compared with 100-150 m^3 ten years
ago). It will be necessary to adopt the recently developed
kraft cooking process now in use at a few mills followed be
treatment with oxygen, ozone, oxygen and hydrogen peroxide.
An advanced cumputerized system is also necessary.

The bleaching of high-yield pulp without removing the
lignin is presently attainable through hydrogen peroxide
treatment. Oxygen is likely to assume a prominent place in
high-yield bleaching as more is learned about how to control
secondary radicals such as hydroxyl radicals and superoxide
anion radicals.

The kraft cooking process utilizes sodium hydroxide and
sodium sulphide as base chemicals. Certain interests are
working to eliminate sulphur from the cooking process. The
Naco process, based on sodium carbonate and oxygen, has been
adopted by two mills in Italy. The use of broadleaf species
(e.g. aspen and poplar), straw and recycled fibers has
confirmed in practice what was known for years within the
research fraternity: namely, that a soda-oxygen process can
replace kraft cooking in certain applications. There is
undoubtedly great potential for increased use of oxygen in
this sector of the pulp industry.

6 CONCLUSION

The use of oxygen, ozone and peroxide has played, and will
continue to play, a vital role in transforming the pulp
industry to an industry of the future, characterized by
efficient production of high-quality products and operating
in harmony with nature.

REFERENCES

1.   J.Gierer, Holzforschung 44 (1990) 387-400.
2.   S.Steeken, J. Chem.Soc. Faraday Trans. 83 (1987) 113-
     124.
3.   G.V. Buxton,J.of Physical and Chemical Reference Data
     17:2 (1988) 513-886.

Uses of Hydrogen Peroxide in Oxidizing Processes Industrial Applications

J. P. Schirmann
ELF ATOCHEM, LA DEFENSE 10, 92091, PARIS, FRANCE

Everybody knows the common ways hydrogen peroxide has been used over the years. Everybody might know also that the bubbles evolving from household hydrogen peroxide are made of oxygen.

It is true that hydrogen peroxide might thus appear to people having received an education in chemistry, as a very convenient way for the storage of oxygen : liquid, stable, it decomposes also very easily at room temperature into water and oxygen.

$$H_2O_2 \xrightarrow{\text{Cat}} H_2O + \frac{1}{2} O_2 \qquad (1)$$

This reaction requires only a tiny quantity of a well chosen catalyst to obtain a smooth evolution of gas.

However such an application of hydrogen peroxide is scarce as it is an expensive source of molecular oxygen as one might easily understand, looking at the reaction of synthesis of hydrogen peroxide.

$$H_2 + O_2 \longrightarrow H_2O_2 \qquad (2)$$

compared to the reaction of decomposition (1). Only half of the oxygen engaged can be recovered, and the cost of the storage, using hydrogen, does not urge chemists to use this valuable chemical as a source of oxygen.

Why is hydrogen peroxide so largely used in various industries ? Why does it continue to be a stellar performer in chemical trade with a growth rate approaching 10% ? New facilities continue to be built all around the world.

The answer is that, depending on conditions of use, hydrogen peroxyde is a very versatile and effective oxidizing agent as a source of what is called "active oxygen" compared to "molecular oxygen" resulting from simple decomposition :

$$H_2O_2 \longrightarrow H_2O + \text{"O"}$$
$$H_2O_2 \longrightarrow H_2O + \frac{1}{2} O_2$$

Where does this versatility come from ? It comes from the fact that this symetrical molecule exhibits a wide variety of mechanisms of action.

It is clear that the molecule H_2O_2 cannot react as such. Formally one has first to break one of the three bonds of this molecule :

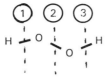

One can break one of the two O-H bonds and then the so called "peroxidic" linkage O-O is conserved in what is called perhydroxyl group OOH.

Or the peroxidic linkage O-O can be broken down and in this case the active intermediates formed will be only hydroxyl groups OH.

- SCHEME   A -

Depending on the fact whether the electron pair of the broken bond is shared or not by the two new entities , the reaction sequence will involve either an ionic or a free radical pathway as shown in Scheme A.

This first reaction act is what is called the "activation of hydrogen peroxide". This activation might be just natural as H_2O_2 exhibits a weakly acidic character in aqueous solutions.

$$pKa = 10^{-6}$$

and it is obvious that depending of the pH of the medium, the chemist will be able to generate even in huge concentrations, various ionic species having either a marked nucleophilic character (pH > 7) or a marked electrophilic character (pH < 7).

In alkaline aqueous medium, hydrogen peroxide reacts with the hydroxyl anions OH$^{\ominus}$ to give perhydroxyl anions HOO$^{\ominus}$ according to the equilibrium.

This equilibrium is shifted to the right when the basicity of the medium increases. This emphasizes the nucleophilic character of the peroxidic linkage O-O, through the formation of the HOO$^{\ominus}$ anion.

This perhydroxyl anion HOO$^{\ominus}$ is considered as a supernucleophile as its reactivity towards electrophilic entities is about two hundred times higher than that of the HO$^{\ominus}$ anion.

For example in the absence of any other reagent, the HOO$^{\ominus}$ anion is able to oxidize an other molecule of hydrogen peroxide, via an unstable transition complex which decomposes with release of molecular oxygen.

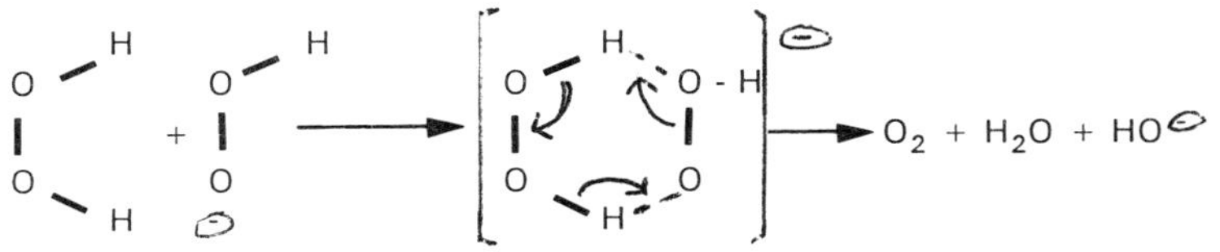

This reaction is responsible of the well know instability of hydrogen peroxide in alkaline medium.

But this decomposition reaction of H_2O_2 can be largely or completely overridden in the presence of an electrophilic substrate. One of the main industrial application is the synthesis of aliphatic amine-oxides, which are used in detergency, according to the scheme :

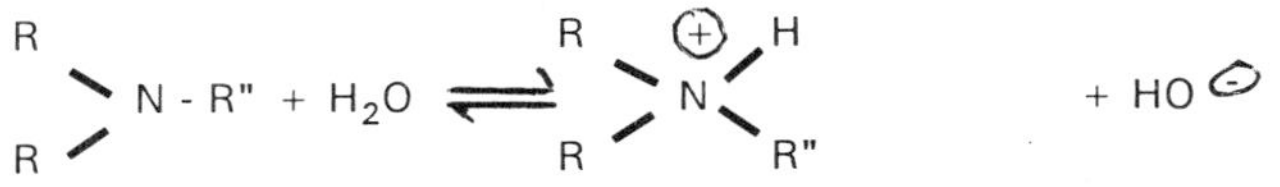

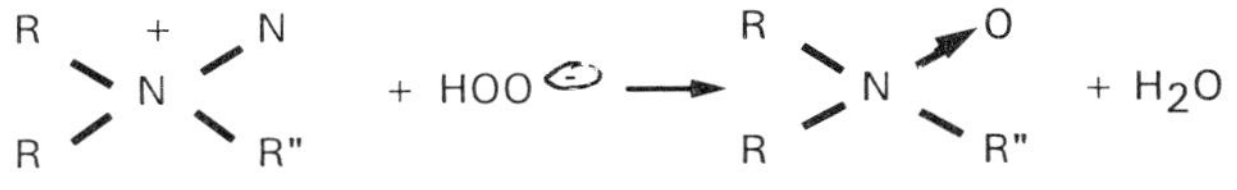

Aminohydroperoxides and aminoperoxides can also be formed very easily at room temperature starting from ammonia, H_2O_2 and a carbonyl derivative :

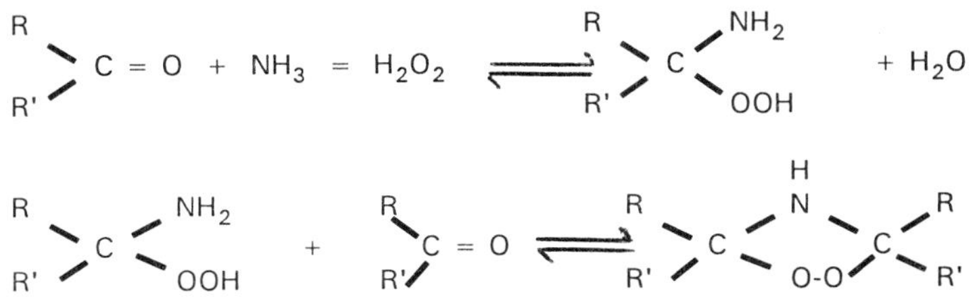

or higher peroxides.

The pyrolysis of such cyclic peroxides leads to a rearrangement which, in the case of cyclohexone, is correctly realized in an industrial process to manufacture 12-amino dodecanoic acid.

Another important application of HOO$\ominus$ is the degradation of lignin in the bleaching of chemical and mechanical pulps.

In acidic medium, H_2O_2 is much more stable as the equilibrium

$$H - O \diagdown_{O} - H \rightleftharpoons HOO\ominus + H\oplus$$

is completely shifted to the left. The formation of an oxonium structure is even observed, resulting from the solvation of protons H$\overset{\frown}{\oplus}$ by hydrogen peroxide.

$$H - O\diagdown_{O^{\diagup}}H + H\oplus \rightleftharpoons H \quad O\diagdown_{O}\diagdown^{H}_{\oplus^{\diagdown}H}$$

This structure changes the nature of the peroxide bond and thus increases its stability.

But when the reaction medium contains a substrate which is a more powerful nucleophile than H_2O_2 (acid, alcohol, ketone, oxometal group) protonation occurs on the latter and a new electrophilic intermediate is generated and reacts with H_2O_2.

This pathway is a very important one in the chemistry of hydrogen peroxide. The peroxidic linkage O - O is conserved and transferred into a new disymmetric molecule which is called a "peroxycompound".

It can be illustrated by the mechanism of formation of peroxyacids from acids, hydroperoxides from olefins or alcohols and peroxides from ketones or acyl chlorides.

Such procedures are carried out on large scale in industry for the manufacture of performic acid or peracetic acid which are strong oxidizing agents of olefins leading to various kind of epoxides (soybean oil, linseed oil, specialty epoxies).

For example, peracetic acid is formed following the scheme :

$$CH_3 - C\diagup^{O}_{OH} + H\oplus \rightleftharpoons CH_3 - C\diagup^{OH}_{OH}\oplus$$

$$CH_3\diagdown_{CH_3}C\overset{OH}{\underset{OH}{\diagup}}\oplus + \ominus O - OH \longrightarrow CH_3\diagdown_{CH_3}C\overset{OH}{\underset{OH}{=}}OOH$$

$$CH_3 - C\overset{OH}{\underset{OH}{-}}OOH \rightleftharpoons CH_3 - C\diagup^{O}_{OOH} + H_2O$$

and is used in the epoxidation of soybean oil (plasticizer of PVC) and in dairy desinfection.

ter-Butyl hydroperoxide is also readily formed starting from H_2O_2 and isobutene in acidic medium.

$$\begin{array}{c} CH_3 \\ \diagdown \\ \diagup \\ CH_3 \end{array} C = CH_2 + H^{\oplus} \longrightarrow \begin{array}{c} CH_3 \\ \diagdown \\ \diagup \\ CH_3 \end{array} C^{\oplus} + {}^{\ominus}O\text{-}OH \rightarrow \begin{array}{c} CH_3 \\ | \\ CH_3 - C - OOH \\ | \\ CH_3 \end{array}$$

and is used as a precursor of di terbutylperoxide useful as polymerisation catalyst. Many others peroxides used in polymers manufacture are synthesized starting from H_2O_2 and ketones.

In all these activation processes, the oxidizing properties of H_2O_2 have been simply transferred to another molecule acting as a support.

Generally speaking, hydrogen peroxide is a <u>general reagent</u> for the introduction of a peroxidic bond into a large number of molecules.

In this transfer, the oxidizing power might be slightly decreased, but most of the time markedly increased, depending of the structure of the new peroxidic compound formed. In general the reactivity increases in the order :

$$R\text{-}OOH < HOOH << R\text{-}C \overset{\displaystyle \nearrow O}{\underset{\displaystyle \searrow OOH}{}}$$

Within the percarboxylic acids serie, we observe the sequence

$$CH_3 < C_6H_5 < m\text{-}Cl\, C_6H_5 < H < CF_3$$

Other industrial applications of percarboxylic acids oxidation are the production of caprolactone, by the Bayer Villiger reaction :

and the ELF ATOCHEM hydrazine process :

$$2NH_3 + H_2O_2 + 2 \begin{array}{c} R \\ \diagdown \\ \diagup \\ R' \end{array} C = O \xrightarrow{\text{Act.}} \begin{array}{c} R \\ \diagdown \\ \diagup \\ R' \end{array} C = N\text{-}N = C \begin{array}{c} R \\ \diagup \\ \diagdown \\ R' \end{array} + 4H_2O$$

$$\begin{array}{c} R \\ \diagdown \\ \diagup \\ R' \end{array} C = N\text{-}N = C \begin{array}{c} R \\ \diagup \\ \diagdown \\ R' \end{array} + 2H2O \xrightarrow{\Delta} N_2H_4 + 2 \begin{array}{c} R \\ \diagdown \\ \diagup \\ R' \end{array} C = O$$

$$\overline{2NH_3 + H_2O_2 \longrightarrow N_2H_4 + H_2O}$$

which is run in France and in Japan.

Hydrazine is used as a rocket fuel and for the manufacture of azo compounds (polymerisation catalysts, porophore agents) and herbicides.

In strongly acidic medium, the oxonium ion formed by the solvation of the proton H $\oplus$ can be split into H_2O and the cation HO $\oplus$

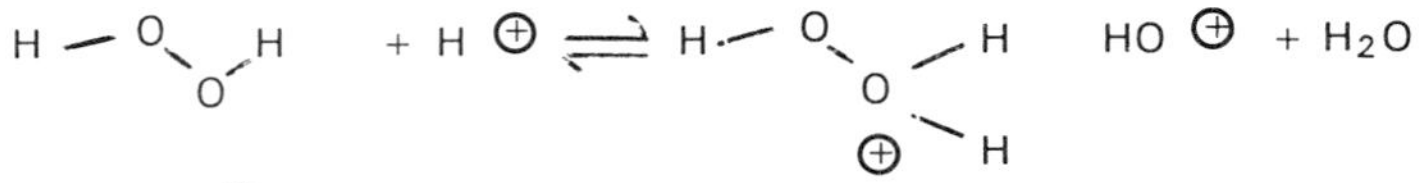

The cation OH $\oplus$ exhibits a very strong electrophilic character towards aromatics, alcohols or even alkanes.

The most important industrial application of this reaction is the hydraquinone-pyrocatechol process developed by Rhone-Poulenc.

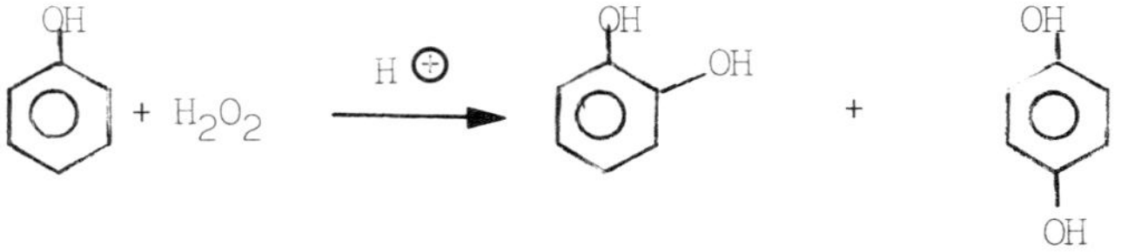

Catalytic oxidations of all kind of substrates have been the subject of many papers in the past fifty years. Most of them deal with homogeneous catalysis, particularly metal peroxo compounds obtained from H_2O_2 and many acidic metal oxides like Os O_4, WO_3, MoO_3, Cr_2O_3, V_2O_5, TiO_2, SeO_2, B_2O_3, which activate H_2O_2 via the formation of an inorganic peracid, where the $M = O$ group plays the role of the $C = O$ group of an organic acid.

The principal function of the catalyst is to withdraw electrons from the peroxidic oxygen atoms.

The only industrial application was realized by ENICHEM in Italy in a process in competition with the pyrocatechol/hydroquinone Rhone-Poulenc process :

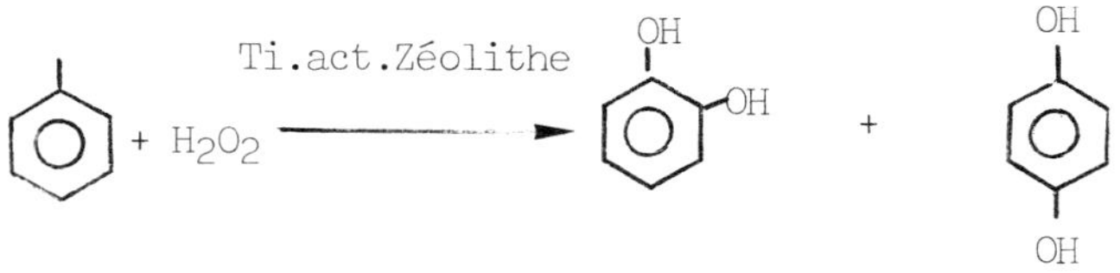

Free radical activation pathways of H_2O_2 is widely used in the polymerization initiation steps. It allows even to obtain functionalized polymers. For example hydroxylated polybutadiene is obtained by polymerization of butadiene initiated with hydrogen peroxide which plays also the role of chain termination agent.

Catalytic generation of HO· radicals is also used for the degradation of organic impurities in waste waters, particularly industrial effluents. The most popular system used is the so called Fenton reagent which can initiate also polymerisation, hydroxylation or oxidative coupling :

$$H_2O_2 + Fe^{2\oplus} \longrightarrow Fe^{3\oplus} + HO^{\ominus} + HO\cdot$$

$$R\text{-}H + HO\cdot \longrightarrow R\cdot + H_2O$$

$$R\cdot + Fe^{3\oplus} \longrightarrow Fe^{2\oplus} + R^{\oplus}$$

$$R^{\oplus} + HO^{\ominus} \longrightarrow R\text{-}OH$$

Hydrogen peroxide is also used in inorganic chemistry for its oxidative and reductive properties. Destruction of cyanide ions in effluents or in gold ores treatment , or reduction of sodium chlorate into sodium chlorite are good examples of industrial applications.

Finally, persalts such as sodium perborate or sodium percarbonate which are solid storage forms of hydrogen peroxide, are widely used as household bleaching powders in Europe and Japan.

More than a potential source of oxygen, hydrogen peroxide is a very versatile, efficient, effective and safe reagent which has no equivalent among the known oxidizers. It is easy to handle and safer to use than many chemicals and new markets are under development for high purity hydrogen peroxide in treating municipal drinking water, or for cleaning and etching phases in semiconductor manufacturing. Treatment of liquid as well as gaseous effluents also looks very promising.

Physical Aspects of The Oxidation of Waste Water

John M. Smith
DEPARTMENT OF CHEMICAL AND PROCESS ENGINEERING,
UNIVERSITY OF SURREY, GUILDFORD GU2 5XH, UK

1.   INTRODUCTION

The biological treatment of waste water presents a major challenge to those of us brought up in the tradition of the process industries on at least three counts. Firstly, by comparison with typical plant throughput the volumes of water to be treated are vast; a city may generate tons per second of waste water. Secondly, biological processes are relatively slow, with the characteristic times running to hours rather than fractions of a second. Finally, although the value of the product may be relatively low that of the raw material is often essentially negative – something has to be done about it.

The field had traditionally been that of the Civil Engineers – Process Engineers have sometimes been heard to describe them as "concrete pourers". But that in fact reflects another aspect of challenge, the investment in existing plants is huge and their depreciation relatively slow. It is the changing social requirements, developing population pressures and increasing land scarcity coupled with the growing awareness of more general environmental requirements which are encouraging the search for compact high capacity designs that can be used either for new installations or as a means of increasing the capacity of existing plant.

2.   THE OXYGEN REQUIREMENT

In the great majority of secondary waste water treatment systems, i.e. after massive suspended solids have been removed by sedimentation, screening or filtration, aerobic biological processes act on organic pollutants and convert them into oxidised end products, usually carbon dioxide or new bacterial cells. The organic material normally provides both the energy and carbon sources for these processes. We will not consider the recent development of alternative anaerobic processes, though in passing it is worth noting that they offer the possibility of integrated energy conservation through the generation of methane fuel gas and the production of significantly less waste biomass – which has to be disposed of – than aerobic systems.

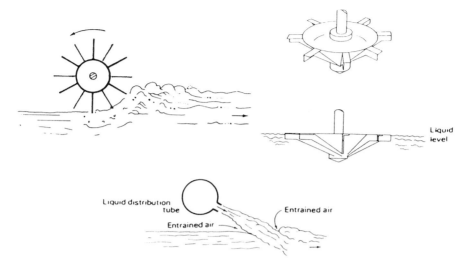

Fig. 1.

The breakdown of organic material by aerobic bacterial organisms requires the supply of oxygen. The quantity of oxygen needed to break down all the biologically degradable pollutants is referred to as the biological oxygen demand (BOD) of the system. Because of the differing rates at which various compounds are broken down by the microbial processes, and possible differences between bacterial cultures taken from various sources, the total oxygen demand may be difficult to determine; standard tests are based on defined culture incubation times, e.g. 5 days. Although since a BOD value is time, pollutant and test culture dependent and so cannot define a pollution level in absolute terms, it has proved a valuable criterion against which to assess both the requirements for the purification of a waste water stream and the quality achieved in the effluent from the treatment process.

The secondary purification systems with which we will be concerned use suspended cultures for the most part. The equipment may range from simple aerated lagoons through oxidation ponds and tanks to advanced bubble column aeration reactors. The bacterial culture is commonly flocculated into masses of the order of 0.1 mm diameter or larger, possibly held together by an extracellular slime. Mechanical or buoyancy induced agitation will keep the flocs in suspension. Although the biological concentration may be low by process industry norms (1 – 5 g/l) the loose structure of the flocs may raise the solids volume fraction to levels where the bulk viscosity is significantly raised.

The treatment of waste water imposes almost unique process operating limitations, centred on the quantities involved and their variability – both as between different locations and in time at a given site. This has led to considerable diversity in process design, with the inevitable feeling that there is no "best" approach. The universal availability of atmospheric oxygen has been exploited in a variety of surface thrashing aerators, plunging jets and waterfalls, arranged in various types of lagoon or circulating ditch, (Fig. 1). In recent times more compact installations have used enriched air or pure oxygen introduced in intensive mixers – venturi disperses or by submerged turbine agitators. These have proved to be particularly effective in dealing with the concentrated effluents from industrial processes and in retrofitting existing installations to upgrade the quality and or volume performance. Other installations exploit hydrostatic or applied pressure to increase the mass transfer rates. At least one objective of these intensified processes has been a reduction in the ground area required.

The very great volume of most waste-water streams sets a premium on the scale of the equipment and on the operating costs – the largest factor in which is usually the energy consumption.

In order that that the culture remains healthy it is obviously necessary to ensure that there is sufficient oxygen supplied to the system. The culture is likely to suffer if anoxic conditions develop; the danger of the system moving unintentionally into an anaerobic regime must be avoided. The physical processes involved as the oxygen is transferred from the gas phase through the liquid to the organism are represented in Fig. 2. Each stage in this chain of transfer processes is complex. Amongst the effects omitted from this description are the influence of surface contamination at the air/water interface, and the reduction in oxygen transfer consequent on any reverse diffusion of metabolic products such as CO_2. In waste water treatment systems the rate limiting factors are usually those connected with the gas to liquid phase transfer.

Reducing the mass transfer model to its simplest form we have

Fig. 2.   The Oxygen Transfer Processes .

a)   Gas compression
b)   Gas distribution
c)   Oxygen transfer at bubble surface
d)   Gas dispersion and recoalescence
e)   Transport through bulk liquid
f)   Transport to floc surface
g)   Transport within floc
h)   Transfer through cell wall
i)   Diffusion within cell to location of metabolism.

$$N_A = k * A * \Delta C$$ (1)

in which N_A is the total rate of transfer of the desired species, i.e. oxygen, (mol/sec), k is the overall mass transfer coefficient $(mol/sec)/(m^2.mol/m^3) = m/s$, A is the available contact area (m^2) and ΔC is the concentration driving force between the two bulk phases harmonised in terms of the equivalent saturation value of the alternate phase. The mass transfer coefficients each side of the interface depend on the local hydrodynamic conditions – most especially the turbulence levels. The contact area A reflects the local bubble size distribution, and likewise is dependent on the intensity of agitation.

Since turbulence affects both the mass transfer coefficient and the surface area advantageously, it is convenient to combine the k and A in the equation in terms that reflect the ability of the system to transfer mass on a volumetric basis: A can be replaced by a $\equiv$ A/V (m^{-1}), where V is the volume of liquid in the system and, in systems where the resistance on one side of the interface far exceeds that on the other, the k by either k_L, the liquid side coefficient or k_G the gas side coefficient. When a fast reaction takes place in the liquid phase, (as for example when, as is common in testing waste water treatment systems sodium sulphate is added to reduce the dissolved oxygen concentration essentially to zero), the liquid side resistance to mass transfer is negligible relative to that arising from the multicomponent diffusion within the gas bubbles. In this case we can write:

$$Na = (k_G a) \cdot \Delta c \cdot V \qquad (2)$$

In the more common case of physical absorption without a reaction, the resistance to mass transfer mainly arises in the liquid phase. Diffusional resistances are much lower in gases than in liquids. The corresponding expression is:

$$Na = (k_L a) \cdot \Delta c \cdot V \qquad (3)$$

These expressions are commonly used in chemical reaction engineering, when the volume of the reactor is well defined. There is something of a difficulty in applying the concepts (and hence the established CRE correlations) to water treatment since the gas to liquid mass transfer action is usually completed within a relatively small fraction of the containing vessel or basin.

Whereas the gas and liquid side mass transfer coefficients are quite difficult to influence during process operation – the minimum turbulent eddy size varies only as the one fourth root of the local energy dissipation rate for example – the contact area, reflecting as it does the bubble size distribution – and the driving force can be influenced to a degree by factors within the control of the designer and operator.

3.   CONTACT AREA CREATION

The interface between gas and liquid can be generated or increased in a multitude of ways. A broad distinction can be drawn between devices that disperse the water as sprays or fountains, including those surface aerators that pick up water from a free surface and project drops outwards, and those that work mainly in the bubble regime. Bubbly systems, almost by definition, have to provide energy to compress the air, though this need not always be the obvious mechanical compression but may, as with jets and ejectors, utilise the momentum of a liquid flow to achieve the desired result.

Although it might appear that there are advantages accruing from reduced rising velocity and a greater contact area per volume of gas supplied by generating the smallest possible bubbles, interfacial contamination builds up rapidly on fine bubbles (<0.1mm). Furthermore bubbles less than about 2mm diameter rise rectilinearly and generate very little turbulence in the surrounding liquid. The resulting poor mixing is not conducive to effective mass transfer.

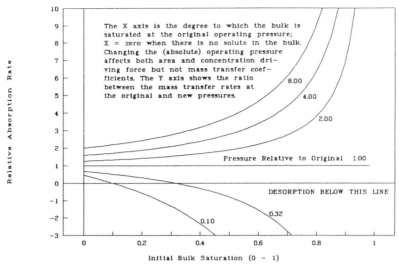

Fig. 3.   Absorption as a function of saturation and changing pressure

4.   CONCENTRATION DRIVING FORCE

The solubility of atmospheric oxygen in water is about 9 g/m^3 at ambient temperature and pressure. In equilibrium with pure oxygen the solubility is about five times this. A aerobic system will only be viable if the dissolved oxygen concentration is not zero, so that the driving force is always less than the difference between zero and these solubilities.

An important group of bubble based oxidation devices use hydrostatic pressure to increase the driving force for mass transfer. Barometric columns have been studied for several years (e.g. Jackson)[1] but the most spectacular applications to waste water treatment have been those involving the Deep Shaft, which will be discussed later. It is however relevant to consider here the effect of ambient pressure on mass transfer from a bubble. We can consider a bubble in the 3 to 5 mm size range, these have a rise velocity almost independent of diameter and so can be expected to be associated with relatively unchanging mass transfer coefficients.

The relative mass transfer rate as the pressure changes can be expressed in terms of the equilibrium conditions. Three factors are involved, the difference between external concentration (usually not zero) and the internal concentration – which varies with the reciprocal of the cube of the diameter of the bubble and the surface area, varying with the square of the bubble diameter. The relative absorption rate follows the curves of Fig.3. Amongst other things this figure shows that very small bubbles, as may be produced by cavitation, can lead to desorption from the bulk fluid.

5.   SYSTEM PURITY

Waste water systems are by definition not pure. Solutes and suspended solids have profound effects on bubble dynamics and mass transfer processes. It has become normal – if unscientific – to lump together these effects in practical waste water systems in terms of a so-called α factor, the ratio between mass transfer rate in the contaminated systems to that which would be achieved with similar operating conditions in a clean system. It is salutary to remind ourselves how the presence of contamination can severely influence mass transfer results.[2,3]

1. Hydrophobic contaminants e.g. oils and many antifoams will naturally accumulate on the gas/liquid interface. As little as 20ppm silicone oils reduce the mass transfer factor to about one third of the value in clean systems. These oils constitute an additional layer of resistance through which oxygen must diffuse. Additionally they will tend to constrain the gas/liquid interface so that it becomes rigid, reducing the rise velocity of the bubble and limiting internal circulation and exterior mixing.

2. Soluble surfactants, e.g. detergents, have a considerable influence. Changes in surface tension directly affect coalescence rates and bubble size distribution.

3. Soluble polar organic compounds – especially higher alcohols – produce dramatic effects on bubble size distribution as a result of coalescence repression. This polarity may influence coalescence (or dispersion) rates as a result of modification to the charge distribution within the heterogeneous system.

4. Ionic materials likewise modify the electrostatic charges in the dispersion. This is most noticeable in terms of the phenomena associated with coalescence repression.

Industrial and domestic waste waters may contain many active materials. Effects are not usually simply additive so the tradition has been to work in terms of the α factor for a particular system.

Much quantitative evaluation of waste water treatment systems has been clouded by the attempt to measure mass transfer rates in oxygen free systems. This starting condition is usually generated by adding some sodium sulphite to the system to remove the dissolved oxygen. Unfortunately even small amounts of ionic salts modify coalescence rates and hence the local specific area available for mass transfer. The reducing effect of the sulphite is usually catalysed and there has been considerable confusion about the kinetics of this process. Finally since a continuing reaction in the liquid phase will result in the k_G being measured rather than the k_L the results are of doubtful relevance to most waste water applications.

At least in laboratory studies it is often possible to work in systems alternately oxygenated by air or purged with nitrogen. Other methods used include modifying driving force by changes in ambient pressure. Oxygen solution can be followed in well mixed systems by fast response (< 1 sec) dissolved oxygen probes, though the interpretation of data from large equipment requires assumptions about both liquid and gas mixing patterns. In recent times there has been considerable success in the analysis of the performance of closed systems in terms of overall inlet and outlet gas analyses coupled to dissolved oxygen measurements in

the liquid phase.   The catalysed hydrazine system has also been useful in
providing useful performance data.

6.   ENTRAINMENT AND DISPERSION - PLUNGING JETS

As every small boy knows, a liquid jet plunging into a pool entraps
air.   Van de Sande[4] and van de Donk[5] have studied this phenomenon, and
large scale application is made in the effluent plant at the DSM
factories near Maastricht.   The air is taken down in a biphasic plume
until the velocity of the submerged jet is reduced to the point at which
bubbles can accumulate, coalesce and escape from the system, Fig. 4.
Entrainment is a function of jet length diameter and velocity and such
properties as the roughness of the jet have considerable influence.   The
DSM installation is particularly interesting; it uses multiple inclined
jets to purify the entire works effluent from the petrochemical and
ammonia plant at Geleen.   The optimal design for jet systems leads to
impact velocities in the range of 2-4 m s^{-1} with jets in the range 2-5 cm
diameter - the minimum size being set by blockage and similar process
constraints.   High velocity small jets can be more efficient though the
volumetric capacity limitation are likely to be limiting.   The best
conditions come near to those of simple overflow waterfalls with a
minimum upstream rise of liquid level (ca 0.5m) a specification which
would lend itself to widespread application in the third world.   The DSM
plant is particularly interesting since by control of the local aeration
rate and balancing a combination of domestic sewage and the ammonia plant
effluent it has been possible to achieve successful denitrification
performance in a very large scale

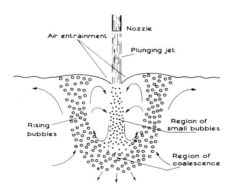

Fig. 4.   The operation of a plunging jet.

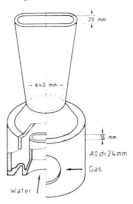

Fig. 5.   The Bayer Slot Aerator

installation.   The process is reasonably efficient, giving plant performance in the range 1 – 2 kg O_2 per kWh with jets of 5 cm or so diameter.   Since the optimum impact velocities are modest, corresponding to a free fall of half a metre or so, water treatment technology based on this principle would appear to be attractive in developing countries, using simple high volume low head pumping devices.

7.   EJECTORS

One way of considering a plunging jet is as an ejector without a housing.   Ejectors (and forced flow ejector devices) give additional control to the gas–liquid dispersion and mixing process.   Amongst the most successful designs used for waste water treatment are those of Zlokarnik[6] which uses a slot configuration, the basic philosophy of which is aimed at separating the bubbles in the generated bubble swarms as rapidly as possible in order to reduce the probability of consequent coalescence, (Fig.5).

Bubble generation from a submerged diffuser is simple and widely applied, though the use of sintered or finely perforated plates in waste water systems is hazardous because of the danger of bacteriological fouling of the dispersing surface.   This problem is to some extent alleviated by using diffusers made of polymer foam.   The natural elasticity of the material allows partial blockages to be expelled before they become serious.   Some aeration installations use simple pipe discharge surmounted by dispensers to break up large bubbles.   The Kenics system (Fig. 6) is an example of the effective use of this principle.

8.   BUBBLE COLUMNS

The unconfined bubble column generated by a diffuser has two important actions; the oxygen transfer associated with the bubble swarm and a buoyancy induced large scale circulation.

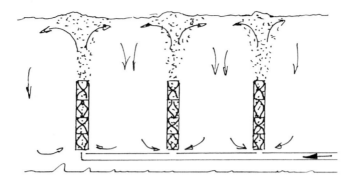

Fig. 6. Kenics Static Mixers used as Air Dispersers

The amount of water brought to the surface by a bubble column is given by
the simple dimensionless equation[7]

$$\frac{Q_w^3}{Q_G \cdot g \cdot H^5} = 4 * 10^{-4} \qquad (4)$$

The water raised to the surface will circulate the contents of quite
a large basin; providing that there is no excessive density
stratification it will mix to a radius about five times the depth, (Fig.
7). About three circulations is sufficient to mix a tank effectively.
For a depth of 10m and a diameter of 80m a gas flow of 1 m^3/s (212 cfm)
will mix the volume completely in less than 8 hours. Interestingly

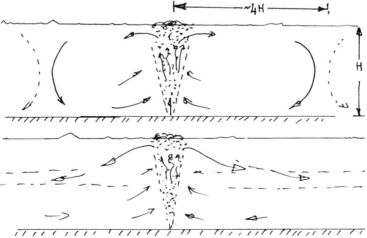

Fig. 7 The Catchment of an Unconfined Bubble Column

enough, except for extreme conditions, the catchment volume of the mixing process is independent of the gas rate or the bubble size distribution, though the circulation and mixing times are directly affected in accordance with the equation quoted.

9.   THE DEEP SHAFT

Buoyancy induced flows are important in another group of aeration installations: those based on loop reactor geometries. The most successful of these is perhaps the Deep Shaft geometry (Fig. 8), developed from knowledge of the behaviour of very tall aerobic fermentation equipment used originally for single cell protein production. The deep shaft consists of a concentric or partitioned shaft, or possible two parallel shafts, possibly of different cross sectional areas. Air injection on one side of the loop induces a rapid circulation (almost inevitably in the order of $1 - 2$ m s^{-1}). Once this is well established air can be slowly introduced perhaps about one third of the way down the downflow leg (downcomer) where the liquid velocity carries the bubbles around the bottom of the shaft so that it can start to provide the buoyancy driving forces in the riser and maintain the circulation.   Below the injection point the gas void fraction in the downcomer is higher than that in the riser (because of the tendency for the gas bubbles to try to rise against the downflow and with the upflow), but the net force balance can maintain the circulation.

The system has several advantages: air need only be mechanically compressed to say 3 bar; the momentum of the water completes the compression almost isothermally and reversibly. The long contact time and high pressure are beneficial for oxygen mass transfer while the low ambient pressure at the top of the shaft should allow effective desorption of CO_2. Floc development in the culture is assisted by the very limited shear stresses in the system. With shaft diameters in the range 3-7 m the volumetric capacity is in the $1-50.10^3$ m^3 range, with

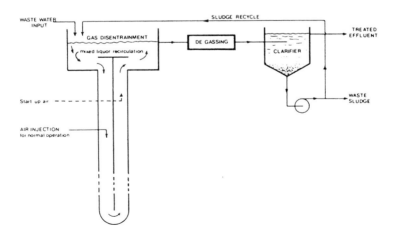

Fig. 8.   The Deep Shaft Principle.

very limited ground area requirements. Given the size of the shaft,
installations are most conveniently built in soft alluvial ground – often
the water level at the free surface can be arranged to be below that of
the natural water table so there is even no danger of contamination of
the surrounding groundwater. There are however three operational
features that must be kept under continuous review[8]. Firstly, the
circulation is inherently unstable – any deviation from steady state can
lead to cyclic instability in the circulation rate: fortunately this can
be countered, at the cost of a small loss of efficiency, by the use of a
small continual bleed of air into the riser. A second problem is that
the flocs tend to retain microbubbles of CO_2 (and N_2?) which may
accumulate and adversely affect the pressure gradients in the downcomer.
The third problem is that the operation depends on the maintenance of a
water-like consistency for the culture suspension. If biological
overload can lead to a mycelial morphology the rheology of the system can
change with disastrous results.

There are other hydrodynamic features of these loop geometries, for
example the radial distribution of gas differs as between riser and
downcomer, leading to differences between the liquid velocity profiles.
Experimental work on ways of introducing gas so as to minimize energy
losses has shown that injection into the throat of a venturi is quite
efficient, with good breakup and thorough distribution of gas throughout
the downstream liquid flow. This principle has been successfully used in
the BOC Vitox process which uses a venturi geometry efficiently to
dissolve oxygen under pressure in a waste water stream. It is often not
realised that introduction of the gas just upstream of the throat often
can be advantageous since the liquid flow there is an elongating flow
which is much more efficient at disintegrating bubbles than a shear flow.
There can be problems with design of the venturi. Boundary layer
separation can occur rather easily in two phase systems, leading to a
much poorer pressure recovery than is obtained in a single phase system.
When this happens the expected advantages of energy efficiency are
largely lost.

The oxygen transfer efficiencies claimed for deep shaft systems are
in the range $3 - 4.5$ kg O_2 per kWh.

The Deep Shaft vividly illustrates the interaction of hydrodynamic
and Biological factors. The flow fields of two phase systems are often
considerably different from those in single phase operations.

10.   SURFACE AERATORS

The many mechanical surface aerators work by thrashing the free
liquid surface, either projecting jets or sprays outwards or driving air
down under the liquid surface, achieved by the interaction of a rotor
with either vertical or horizontal axis, (Fig. 1a,b). In either case
oxygen transfer is limited to the surface region, with little penetration
of bubbles. Deep basins must exploit other circulations to ensure
aeration over the whole depth – some aerators generate large scale
secondary flows at channel bands to provide top to bottom exchange. As
Zlokarnik has shown, in order to maintain performance on scale-up these
aerators should be scaled on the basis of constant Froude Number (N^2D/g)
which leads to excessive power consumption in large installations.

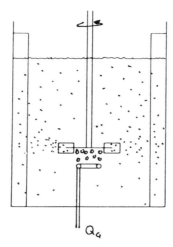

Fig. 9. A typical "standard" submerged turbine arrangement.

11. SUBMERGED TURBINES

In recent years there has been a considerable advance in our understanding of the technology of gas-liquid mass transfer with submerged turbines, at least as applied to chemical reactor engineering. The "standard" reactor configuration uses a six bladed disc, "Rushton", turbine in a tank of H/T rate about 1, (Fig. 9). The impeller diameter is usually between 0.25 and 0.5 that of the tank. Other designs (simple pitched blades or sophisticated hydrofoils as well as disc turbines modified with other disc patterns) have come into use though there are similarities in principle in the way these work. When Rushton impellers

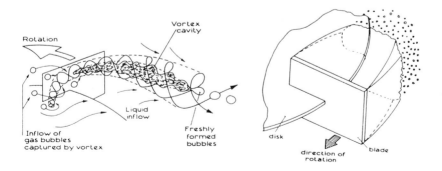

Fig.10 Vortex and large cavities

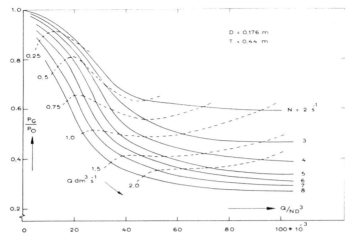

Fig. 11 Power Demand Curve for a Turbine

operate in a gas/liquid system captive ventilated cavities develop behind the blades[9]. At low gas fractions there are spinning vortex cavities in which gas first coalesces and then is dispersed as it splits away from the tail of the cavity. With larger gas fractions the cavities are large and smooth sided, with a different mechanism of dispersion (Fig. 10).

The drag of the impeller is modified as these changing configurations develop and this is reflected in the power demand curve, (Fig. 11). The maximum gas handling capacity is reached at the flooding point, when the buoyancy forces become great enough to overcome the pumping action of the impeller. It has been possible to characterise the

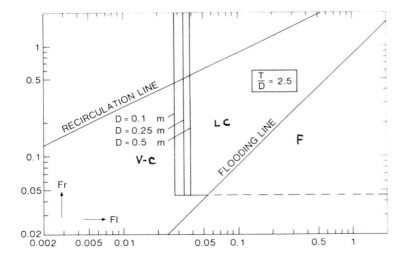

Fig. 12.   Gas–Liquid Flow Map for a Rushton Turbine.

operation of disc turbines in a gas–liquid flow map (Fig. 12), which allows us to determine the flow regime independently of the equipment scale[10]. Performance prediction is most effective when flow regimes are similar: unfortunately the scaling rules lead to the conclusion that most laboratory investigations do not reproduce the conditions of operating plant because of the impossibility of matching all the relevant dimensionless groups simultaneously.

Mass Transfer Performance   The usual predictions of mass transfer performance are dimensional (specified in SI units) and are based on the specific power input, e.g.

$$k_L a = 1.2 \ (P/V)^{0.7} . v_s^{0.6} \tag{5}$$

in which V refers to the volume of the vessel in m^3 and v_s the gas superficial rate m s^{-1}, based on the total vessel cross sectional area. To relate this to an open basin requires a different basis for calculation. One approach is to base the effective volume on the impeller diameter rather than on the vessel as a whole. Arguing that a typical reactor uses an impeller with a diameter one third of that of the vessel leads to the modified equation:

$$k_L A \ = \ 0.93 * P^{0.7} * D^{-0.3} * Q_G^{0.6} \tag{6}$$

in which P is the power (W) supplied to the impeller of diameter D (m) through the shaft whilst gas is being supplied at a rate of Q_G (m^3 s^{-2}). From this relationship it is easy to show that the power efficiency for physical absorption by a turbine cannot be greater than 1.3 kg O_2 per kWh.

12.   CONCLUSION

This paper has considered factors affecting the transfer of oxygen to waste water in the context of various types of process equipment. The energy involved in the two activities of gas compression and generation of contact surface provides the major operating costs. It is through improved understanding of the multiphase processes involved that future process developments can be expected.

NOTATION

A	Contact area for mass transfer	m^2
a	Specific contact area, m^2/m^3	m^{-1}
ΔC	Concentration driving force	$mol \ m^{-3}$
D	Impeller diameter	m
g	gravitational acceleration	$m \ s^{-2}$
H	Vessel or pool depth	m
k	overall mass transfer coefficient	$m \ s^{-1}$
k_G	mass transfer coefficient, gas film basis	$m \ s^{-1}$
k_L	mass transfer coefficient, liquid film basis	$m \ s^{-1}$
N	impeller speed	s^{-1}
N_A	Mass transfer rate	$mol \ s^{-1}$
V	Reactor (liquid) volume	m^3
P	Power input	W

Q_w Volume rate, water at pool surface $m^3 s^{-1}$
Q_G Volume rate of gas injection $m^3 s^{-1}$
T Vessel diameter m
V Liquid volume in reactor m^3
v_s Gas superficial velocity, $m^3 s^{-1}/m^2$ $m s^{-1}$
α Correction factor, effect of contamination $-$
ρ Liquid density $kg\ m^{-3}$
μ Liquid viscosity $Pa\ s$
Fl_G Gas Flow number $\equiv Q_G/ND^3$ $-$
Fr Froude Number $\equiv N^2D/g$ $-$

REFERENCES

1. Jackson M L and Shen C-C
    Aeration and Mixing in Deep Tank Fermentation Systems
    AIChE Jnl (1978) 24, 63 – 71

2. Warmoeskerken M Smith J M
    Surface Contamination effects in Stirred Tank Reactors,
    Int. Sympos. Mixing, Mons, Belgium. 1978,   Paper C-13 pp 1-16

3. Zlokarnik M.
    Sorption Characteristics ....
    Advances in Biochemical Engg. 8, (1978) 133 –151

4. van der Sande E and Smith J M
    Mass transfer with plunging Jets
    Chem. Eng. Jnl, (1975), 10, 225 – 233

5. van de Donk J, van der Lans R and Smith J M
    The effect of contaminants with a plunging jet contactor
    Proc. 3rd Eur. Conf. on Mixing, BHRA., (1979) 289 – 3902

6. Zlokarnik M.
    Tower Reactors for Waste Water Treatment, in
    Rehm & Reed, Biotechnology, vol 2,   VCH (1985)

7. Smith J M and Goossens, L
    The Mixing of Ponds with Bubble Columns
    Proc. 4th Eur. Conf. on Mixing, BHRA., (1982) 71 – 80

8. Smith J M and van der Lans R.
    Liquid Circulation in a Bubble Column Loop
    Proc. 4th Eur. Conf. on Mixing, BHRA., (1982) 471 –476

9. Bruijn W, van't Riet K and Smith J M
    Power Consumption with Aerated Rushton Turbines,
    Trans. Inst. Chem. Eng., (1974), 52, 88 – 104

10. Warmoeskerken M M C G
    Gas dispersing characteristics of Turbine Agitators
    Doctoral Thesis, Delft, 1986